*The Problem with Solutions*

# The Problem with Solutions

WHY SILICON VALLEY CAN'T HACK
THE FUTURE OF FOOD

Julie Guthman

UNIVERSITY OF CALIFORNIA PRESS

University of California Press
Oakland, California

© 2024 by Julie Guthman

Library of Congress Cataloging-in-Publication Data

Names: Guthman, Julie, author.
Title: The problem with solutions : why Silicon Valley can't hack the future of food / Julie Guthman.
Description: Oakland, California : University of California Press, [2024] | Includes bibliographical references and index.
Identifiers: LCCN 2023054098 (print) | LCCN 2023054099 (ebook) | ISBN 9780520402669 (hardback) | ISBN 9780520402676 (paperback) | ISBN 9780520402683 (ebook)
Subjects: LCSH: Agricultural innovations—California—Santa Clara Valley (Santa Clara County) | Food—Technological innovations—California—Santa Clara Valley (Santa Clara County) | Food industry and trade—Technological innovations—California—Santa Clara Valley (Santa Clara County) | High technology industries—California—Santa Clara Valley (Santa Clara County)
Classification: LCC S494.5.I5 .G87 2024 (print) | LCC S494.5.I5 (ebook) | DDC 338.1/6097946—dc23/eng/20240213
LC record available at https://lccn.loc.gov/2023054098
LC ebook record available at https://lccn.loc.gov/2023054099

Manufactured in the United States of America

33  32  31  30  29  28  27  26  25  24
10  9  8  7  6  5  4  3  2  1

# Contents

Preface *vii*

Introduction: The Origins of Solutions   1

1  Silicon Valley and the Urge to Make the World a Better Place   29

2  Agrifood Solutions before Silicon Valley   54

3  Silicon Valley Bites Off Agriculture and Food   74

4  Alternative Protein and the Nothing Burger of the Techno-Fix   96

5  Digital Technologies and Plowing Through to the Problem   118

6  Silicon Valley Thinking Comes to the University   138

7  Big Ideas and Making Silicon Valley–Style Solution-Makers   159

Conclusion: The Pessimism of Solutions and the (Cautious) Optimism of Response   178

Acknowledgments *195*
Glossary of Terms *199*
Notes *203*
Bibliography *223*
Index *241*

# *Preface*

In 2016 I attended an event, held in a city winery, in San Francisco's South of Market district. In cooperation with Rabobank, a Dutch financial institution that claims more than one hundred years of investments in the food and agriculture industries, an outfit called FoodBytes! had sponsored the event, the fourth of many it held until the COVID-19 pandemic hit in 2020. FoodBytes!, according to its website, strives to connect food industry leaders and investors with start-up companies that are "innovating and disrupting the food chain with groundbreaking ideas in food, agribusiness and technology." Four hundred people attended the event, including start-up entrepreneurs, representatives of conventional agribusiness firms, venture capitalists, foodies, and the well-paid techies who live in this definitely gentrified neighborhood.

All of us were packed into a very tight space. The main room of the venue contained about a dozen vendor tables for showcasing innovations and handing out samples of those innovations that came in edible form, along with a makeshift stage where the main attraction took place. A side room provided more private space for individual start-ups to meet with venture capitalists, should the former be so lucky to receive a bite of interest. For the first part of the event attendees were plied with wine and encouraged to network and view the displays. The host then welcomed "all of you who care about food and who care about innovation" to be seated for the main event. It began with about ten sixty-second lightning pitches, followed by a "fireside chat." Without, thankfully, a fireplace in the vicinity on this very hot day, the panel featured representatives from Rabobank and a food-oriented "accelerator" that mentors and makes connections for start-ups wanting to take the next steps. Panelists' comments were geared toward ginning up enthusiasm for the agrifood "space."

Then came the highlight: twelve curated five-minute pitches each followed by five minutes of Q&A from the judges and audience, closing with talks by successful alumni of FoodBytes! pitch events. During the alumni talks, attendees voted on their favored innovation; the winner of the vote received a prize, as did the judges' choice. Centered on the day's themes of nutrition, waste, and efficiency, the featured products ranged from plant-based shrimp, cookies made with cricket flour, and solar panels for greenhouse production, to data systems for reducing resource use in agriculture and a sharing economy marketplace of farm equipment ("an Airbnb for tractors"). Of the several vendors whose world-changing aspirations most caught my attention, Kuli Kuli Foods proved most riveting. Claiming a solution to food insecurity in the developing world, representatives from Kuli Kuli featured snack bars made from moringa. Moringa tree leaves, they explained, provide a complete protein, containing all nine essential amino acids. And so the snack bar they were offering was a so-called superfood—a food that provides extraordinary nutritional benefits.

Given the nod to food insecurity, I wondered how moringa bars would be distributed and enjoyed by the ever-invoked malnourished children in Africa. It turns out that the Kuli Kuli project was much more than a snack bar and involved a much more elaborate solution. So compelling was the solution, and founder Lisa Curtis's personal story, that Curtis herself spoke as one of the event's featured alumni. A former Peace Corps volunteer, Curtis had visited Niger as part of her service. As a vegetarian, she had often felt sluggish from a diet largely consisting of rice and millet. Luckily, women in the hosting village introduced her to moringa as an antidote to her hunger. Eating it with peanut paste, a mixture the locals called *kuli kuli*, Curtis found her condition immensely improved. Following this experience, she began to tell the villagers—the very people who had introduced her to the moringa plant—to include more moringa in their own diets. Then she hit upon a much bigger idea: to create a women's cooperative that would export moringa to the United States for manufacture and sale in various forms of snacks and supplements. The cooperative would provide a new source of income to the village women, improving their livelihoods and ability to acquire food, while US consumers would get a convenient source of high-protein greens.[1]

After returning to the United States, Curtis quickly raised startup funds from an Indiegogo crowdfunding campaign. Emboldened by her initial success, she expanded her vision to include a project to help reforest Haiti with moringa trees. Not only was the tree drought tolerant; Curtis also hoped to provide Haitian smallholder farmers access to the growing market for moringa leaf powder. This new initiative received support from Whole Foods and the Clinton Foundation as well as another crowdfunding campaign. Eventually Kuli Kuli raised $4.25 million from a venture investment arm of the Kellogg Company, comprising one of the first investments by a Fortune 500 company into a certified Benefit

Corporation (B Corp).[2] By law, most corporations must place shareholder value above all else. B Corps, in contrast, can place social and environmental impact on an equal level with profit, allowing companies certified so to pursue projects of genuine societal improvement. The company's social aims, according to its website, include ending malnutrition, empowering women to achieve gender equality, and planting a tree for each household in the communities where it works.[3]

. . .

Perhaps you're a reader inspired by the Kuli Kuli story and the aforementioned food, agribusiness, and technology event. What I had witnessed that day reflected a burgeoning movement of people who've been immersed in the tech sector and its zeitgeist, seeking to apply tech to transform one of the most intractable yet important domains of planetary existence: the world of food. The problems of food addressed that day—food insecurity, overfishing of the sea, excessive water and pesticide use—are real, and those forwarding solutions were taking concrete steps to address them.

You might have found Curtis particularly elevating. Besides having a great story that demonstrated a long-term commitment to positive transformation, she was passionate about her ambitions, and her excitement was infectious. You might be impressed that she understood that hunger couldn't be addressed by handing out moringa bars (my initial misconception). She recognized that hunger generally stems from poverty and precarious livelihoods and that women disproportionately lack income security. So she had creatively developed a way to enhance the incomes of women in her regions of interest while providing a healthy food in the United

States—a win-win idea if there ever was one. You might also laud Curtis's generous intentions, as indicated in her procuring B corporation status, making the Kuli Kuli plan more than inventive marketing. And you might admire that she had brought along a major food corporation, which would allow her company and the program to scale up and potentially transform many more lives. In a day and age when far too many reasons exist for dissatisfaction, outrage, and despair about our social and ecological worlds, and when you really have no idea how to fix it all, a solution like Kuli Kuli's is really appealing. It is concrete, doable, and for the fixer, emotionally satisfying. A win-win-win.

I suspect that the Kuli Kuli story has enraged another segment of readers, including some of my students. You might see it as an instance of "racial capitalism" in which the poverty of people who have been made poor by capitalist development are now being utilized to sell a commodity to privileged people in the United States (who really don't need what they're buying). You might scoff at Curtis's deal with Kellogg, or Kuli Kuli's becoming a B Corp since you find profit-making and "doing good" an oxymoron. You might find Curtis's intentions "neocolonial," since she was interfering with people who didn't ask for her help and maybe don't want it, and you might find it galling that she was promoting *kuli kuli* to the people who introduced her to it. Perhaps you feel, as I do, that Curtis didn't do her research on the well-documented failures of past initiatives to empower West African women owing to gendered divisions of household income streams and therefore didn't adjust her thinking accordingly.[4]

In a day and age when far too many reasons exist for dissatisfaction, outrage, and despair about our social and ecological worlds, and when you really have no idea how to fix it all, you might bask in this critique, feeling that nothing can be changed until we change it all. Or maybe you're a reader who does not relate to either position, finding them both overdrawn and wondering why the fuss. After all, Lisa Curtis and Kuli Kuli were at least doing *something* to make the world a better place. They at least were working on a solution.

· · ·

As for me, I suppose I've long been somewhat averse to solutions per se. Back when I was nineteen years old or so, I attended the University of California-Santa Cruz (UCSC), known as a hotbed of critical thought to this day. (In an unusual turn of events, many years later I became a professor at UCSC.) I recall coming home from college to visit my family, brimming with ideas and wanting to engage my parents in conversation about what I had learned from reading Marx, Freud, Weber, and Nietzsche—standard pillars of a liberal arts education at the time. I suppose I expected my parents to share my interest, as if they had been waiting on my intellectual adulthood. I was particularly eager to discuss the problems of poverty and inequality stemming from capitalism. But "what is the solution?" my father queried.

I didn't like the question, even though at the time I couldn't articulate why. Why did I have to have a solution to capitalism of all things? Why wasn't thinking critically important enough? What if capitalism, in all its dynamism and complexity, can't be solved in any immediate way? It is not as if six hundred or so years ago, when feudalism was in crisis, someone came along and said: "Ah, I have a solution: capitalism. Here is my plan for replacing feudalism with capitalism and my vision of how the latter will develop over the next six hundred years."

One way to read my father's query was that he was pushing me about the failed twentieth-century experiments in state socialism (another conversation for another book). For this book I offer a different read of his question, one that is more relevant to today: underlying the question "What is the solution?" is the idea that if you're going to think about or engage with a problem, you need to aim to solve it. And you need to aim to solve it in immediate and doable terms or you won't succeed. The problem with this view, besides squashing all additional conversation and insights a deep dive into a big problem might beget, is that understanding a problem only in solvable terms begins to set limits on the range of possibility of addressing the manifold crises we face in the world.

. . .

What is a solution anyway? Solutions, as described in this book, are finite, narrowly conceived fixes to problems that themselves have been bounded and rendered solvable.[5] They are not Band-Aids, meaning stop-gap measures designed to mitigate an acknowledged larger problem, nor do they make a problem go away for good. Rather, solutions are innovations that purport to address vast and complex problems but are often pursued with limited knowledge of the problem, a narrow set of tools, and a constrained sense of possibility.

To make a solution, that is, you have to choose a problem that is manageable—and you don't necessarily need to understand the problem in its totality. Kuli Kuli proved a solution because it took the problem of malnutrition and made it solvable by trying to create an income stream for women who could sell a crop with promising qualities. But it stopped there and did not consider how many women it could support and whether this income would sustain were moringa-based snack bars to flood the market. Nor did the solution leverage the corporation's ostensible goal to address malnutrition more broadly. Rather than stemming from "radical optimism," as innovators like Curtis describe their outlook, solutions tend to assume that certain things can't be changed.[6]

Although this book largely emphasizes the problems with solutions, I don't want to leave you high and dry with critique alone. My critique of solutions doesn't give you license to sit around and do nothing. We have a world full of oppression and ecological harm that needs addressing. But I want to convince you of a different sensibility and a different approach as you encounter a difficult and deteriorating world. The present moment, especially, needs something much more robust than solutions. A first step in doing social change differently is thus coming to terms with the allure of solutions. This means beginning to let go of, or at least acknowledge, what feels so good about solutions: their immediacy, their finiteness, their manageability. Such qualities feel highly rewarding, even self-validating to the solution-maker but often fail to improve worldly conditions.[7] But embracing critique and only critique won't do, either. You can land in a different place where it's possible to hold and embrace critique about broader systemic problems while engaging in bounded, immediate action in the present. But it takes thinking of change not as a matter of finding solutions but responding instead.

. . .

This book is for those of you who want to tackle one or more major societal problems—for techies, social entrepreneurs, activists, and other changemakers and for young people, especially student readers, who will inhabit these roles in the future. As colleges and universities increasingly encourage you to make a positive impact, many of you will be entertaining choices between the allure of solutions and the challenges of movement building, public action, and political and cultural organizing. I encourage you to avoid the pitfalls of solutions and embrace those other paths and do no less than becoming an engaged world citizen who follows the news and exercises your rights, including the right to vote.

I also aim this book at those who *enable* solutions—financiers, philanthropic funders, university administrators, or nonprofit incubators. You have an important role to play, too, in redirecting action away from the solvable to the responsible and supporting young people eager to bring about a better world. As leaders from all sectors embrace solutions to solve all worldly problems, it is imperative to shed light on other ways of enacting change.

# *Introduction*

The Origins of Solutions

In 2015, with great fanfare, the Massachusetts Institute of Technology (MIT) launched a Media Lab Open Agriculture Initiative, soon to be shorthanded as OpenAg. Per its website, the founders averred that the precursor to a sustainable food system was an "open-source ecosystem of technologies that would enable and promote transparency, networked experimentation, education, and hyper local production." A research collaborative of industry, government, and academia, the aspiration was to use digital technologies to explore and ultimately bring about the future of food, focusing on such areas as urban farming, food transparency and authenticity (meaning knowing where your food comes from), and plant flavor and nutrition.[1]

The open source objective was particularly laudable. It means that the relevant computer code developed by the lab could be made available for anyone to use, copy, or change, possibly with no license fees. Made possible by the lab's home at a university, such arrangements stand in sharp contrast to proprietary technologies emanating from the private sector. Those require users to pay license fees and rarely offer users access to the underlying computer code. Open source technologies are therefore more democratic, providing the possibility for endless tweaking to suit local conditions. Based in a university, the OpenAg lab also created an opportunity for select MIT students to learn and apply engineering skills to address something as meaningful and important as the future of food. The OpenAg Initiative's most high-profile innovation was the "personal food computer," an enclosed chamber the size of a minifridge packed with LEDs, sensors, pumps, fans, control electronics, and a hydroponic tray for growing plants.[2] The Fukushima nuclear disaster in 2011 had allegedly provided the initial inspiration for the prototype. The Initiative's principal investigator, Caleb Harper, had visited Japan shortly following the meltdown of several reactors in a major nuclear power plant, triggered by a magnitude 9.0 earthquake and deadly tsunami. The accident led to widespread radiation of soil, rendering it too toxic for growing food safe to eat. Growing food in soil-less conditions seemed a compelling solution under those circumstances.

Still, the excitement around the personal food computer was based on its digital capabilities. By housing a common platform for farmers to share data and replicate climate formulas from anywhere, the hope was that this platform would enable farmers throughout the world to produce healthy fruits and vegetables, irrespective of climate and seasonality.[3] Such "hyperlocal" produce could drastically reduce the long-distance shipment of produce that has become a mainstay of food globalization. According to some reports, Harper "mesmerized audiences and investors around the globe with a vision of 'nerd farmers' growing Tuscan tomatoes in portable boxes with recipes optimized by machine-learning algorithms."[4] Five years later, however, MIT closed the doors on the much hyped OpenAg Initiative in 2020. Whistleblowers within the lab had reported that the successes were exaggerated, even completely fabricated. The tech journal *Gizmodo* called the prototype a "Theranos for plants," referring to the company founded by Elizabeth Holmes that raised billions of dollars for an unproven technology for blood tests, eventually leading to Holmes's criminal conviction for fraud.[5]

The overhyped personal food computers were not the only cause of OpenAg's demise. Reports that the lab was discharging a fertilizer solution into underground wells led to an investigation and eventual fine by the Massachusetts Department of Environmental Protection. Lab personnel also had solicited major donations from convicted sex offender Jeffrey Epstein, although the donation was never made. Following this spate of embarrassing publicity, MIT opted to shut the entire operation down and dismiss several employees. Former lab researchers later complained that MIT had continued to raise money for the OpenAg lab after they had reported the exaggerated claims to the university.[6]

As fascinating as it is, the MIT OpenAg story interests me less for the palace intrigue than for its illustration of two phenomena: one is the newfound interest among techies to fix agriculture and food with technology—to hack the future of food, as it were—a domain that has experienced a complicated history in relation to technology; the other is the suffusion of Silicon Valley tech culture into the university. Notwithstanding that MIT's very mission is to advance science and technology in service of the public good, OpenAg seems to have adopted more of the Silicon Valley trappings than usual, replete with outsized promises and a particular approach to problem-solving. To both these points, what exactly was the problem that MIT's personal food computer was going to fix? Could desktop-size hydroponic operations really feed the world with healthier food? Could they address the toxic contamination of soils? Was an algorithm the answer to growing delicious Tuscan tomatoes? If so, was that the most pressing need?

Believe it or not, the personal food computer is not really an outlier. Billed as "world changing," many current-day technological solutions to food and agriculture don't solve anything of great importance. Worse, many are not so different than prior agrifood technologies, reproducing many of the conditions and structures innovators putatively aim to disrupt. The main problem with these technological solutions is they don't contend with the inextricable social roots of many of the world's food and agriculture problems— that is, how core resources such as capital (investable money), land, labor, biological material, and expertise are distributed, managed, maintained, and fought over. To the contrary, most solutions are guided by impulses that deliberately skirt these more vexing concerns.

Under less dramatic circumstances and undoubtedly with more integrity, MIT has developed several other programs that converge technological problem-solving with food and agriculture. These include an annual cellular agriculture hackathon backed by the food industry and a Food and Agriculture Club that holds an annual pitch competition. Nor is MIT unique in its food innovation efforts. Colleges and universities from all over the world have jumped on the food and agricultural tech (hereafter "agrifood tech," "ag tech," or "food tech") train, sprouting training programs, research initiatives, student competitions, and more, all geared toward developing technological solutions to the planet's "grand challenges" of food and agriculture: climate change, food insecurity, environmental sustainability, and human health and nutrition. Importantly, much of this activity does not necessarily stem from the traditional agriculture and food science departments but from offices of innovation, engineering departments, private donors, and students themselves. Embracing development and dissemination approaches modeled after Silicon Valley start-ups, all of these efforts seek to have real-world impact during highly fraught times. Even my own university, the University of California–Santa Cruz (UCSC), an institution deeply rooted in the liberal arts tradition and a pioneer in agroecology and alternative food initiatives, is getting in on the ag tech act.

Many of these programs encourage students to be solution-makers in the food and agriculture sector and beyond. Notice the term "solution-maker," connoting something different than activist. Solution-makers seek to *fix a specific problem*, often narrowly defined, whereas activists work with issues and mount campaigns to *change conditions*. One appears straightforward and doable; the other can seem amorphous and difficult. One promises to deliver impact in a relatively short amount of time; the other can seem uncertain and never-ending. One provides near-term emotional gratification that you're doing something; the other requires continued commitment to a larger cause and risks emotional burnout. That the word "solution" shares a word root with "absolution" is instructive. Solutions allow the solver to walk away, as it were, once the solution is implemented. So you can see why solution-making is attractive. But sometimes solutions are not up to the task of changing what needs changing. Indeed, as I argue, *solutions are a product of our present condition, not a response to it*.

## A Lexicon of Solution-Making

Admittedly, it is strange to take issue with solutions. Contemplating solving problems small and big—from how to navigate transportation to work or school, to how to address the crisis of migrants at the US southern border—suffuses human existence. Sometimes a day doesn't go by that I don't hear or use the term. And, of course, you are constantly prevailed upon to imagine solutions to thorny social and political problems. But if you suspend your everyday understanding of solutions, even momentarily, and conceive of solutions as I use them in this book, you might begin to see their limits. For making finite, bounded fixes to problems that themselves have been reduced to the actionable authorizes certain kinds of thought and action while foreclosing others.[7] This becomes apparent through an examination of three distinct impulses, legs of a tripod if you will, that I argue undergird the problem with solutions. Though these impulses are closely related and often intersect, they originate in different contexts. Attending to these different origins will not only allow you to better discern each impulse; parsing them out will also perhaps encourage you to reflect on your own tendencies to embrace one or more or reject them altogether.

Briefly, these three impulses are (1) the techno-fix, or the impulse to address social and ecological problems with technology alone; (2) solutionism, or the impulse to develop solutions in advance of or in disregard of problems; and (3) the will to improve, or the impulse to act on behalf of others for their benefit. In most solution-making situations, all three impulses are present, although generally one is more prominent.

*Techno-Fixes*

Techno-fixes draw on technological advances to solve societal problems, guided by the assumption that humans can engineer their way out of crises big and small, and indeed that creating a technological solution is easier than changing societal structures or people's behaviors. Promoters of techno-fixes want to avoid the messy world of politics that can thwart needed change. Or they want to maintain existing ways of life and alleviate the need to sacrifice the comforts and conveniences of modern-day living. Dating back to the machine age and scientific management in the 1920s, Americans have long valued investments in techno-fixes; indeed, they were a central feature of mid-twentieth-century modernity. One early proponent was Alvin Weinberg, a nuclear engineer who became the head of Oak Ridge National Laboratories in Tennessee. Weinberg championed the idea that engineers could replace social scientists by designing technologies that did not require the more insurmountable goal of having the public change their habits. For Weinberg the key technology was nuclear engineering and the promise of unlimited energy use. The same principles underpinned "better living through chemistry," a variant of the DuPont Company slogan once deployed to convince the public of the positive role chemicals could play in society.[8]

Efforts to engineer the climate are a paradigmatic example of the techno-fix in action today. Although a warming climate undoubtedly stems from the use of fossil fuels to uphold industrial capitalism and a consumerist way of life, sending solar radiation back to space through, for example, ejecting sulfate aerosols into the atmosphere in order to deflect greenhouse gases back into the stratosphere (as if this was simple, desirable, and not potentially catastrophic) allows current ways of life and unfettered capitalism to continue. A contemporary school of environmental thought called ecomodernism explicitly advocates for technologies that allow humans to maintain (or reach) high standards of living with few sacrifices. Ecomodernists continue to advocate for nuclear power as a source of energy as well as technologies of industrial agriculture that "spare land," ostensibly to save large portions of the world for nonhuman nature to exist without human interference.[9]

Techno-fixes are by no means categorically bad. Who among us isn't grateful for the development of solar-powered electric vehicles or doesn't yearn for a techno-fix to the millions of tons of plastic garbage floating in the ocean known as the Great Pacific Garbage Patch? And yet techno-fixes have their limits. An early critic of the principles of the techno-fix was the philosopher Arne Naess, a proponent of "deep ecology." Writing in the 1970s, Naess argued that techno-fixes tend to maintain the status quo, at best requiring no sacrifice and at worse allowing current socioeconomic conditions such as rampant inequality and vested interests to persist.[10] It is easy to see how that is so with geoengineering, which has generally been promoted by people with substantial economic power, indeed many of whom have acquired much of their wealth through fossil fuel–fueled industrialization.[11] Yet techno-fixes can rarely address underlying problems, and sometimes they are completely off the mark. Techno-fixes are by definition narrow solutions, and developers of techno-fixes tend to underestimate "both the scale and complexity of problems and the side effects (or unintended consequences) that such engineered solutions can offer."[12] Worse, techno-fixes can worsen the problem they are trying to fix, while distracting from the conditions that really must be changed. This is in large part because techno-fixes by design ignore the social dimensions of any given problem. As Naomi Klein has passionately argued, addressing climate change really does require us to change how the well-off and comfortable live, and anything different (like a techno-fix) is tantamount to looking away.[13]

The early US response to the COVID-19 pandemic further punctuates these points, as discussed in a September 2020 article by Ed Yong in *The Atlantic*. With the help of former President Trump, he wrote, Americans clung to magical thinking and the promises of quick solutions (such as a vaccine by November 2020, hydroxychloroquine and convalescent plasma, or other medical cures) rather than "navigating a web of solutions, staring down broken systems, and accepting that the pandemic would rage for at least a year"—and "possibly alter our lives forever." Yong went on to note that the problem was not unique to COVID-19. "It's more compelling to hope that drug-resistant bacteria can be beaten with viruses than to stem the overuse of antibiotics, to hack the climate than to curb greenhouse-gas emissions, or to invest in a doomed oceanic plastic-catcher than to reduce the production of waste." Like Naess and Klein, Yong suggested that techno-fixes are unlikely to succeed on their own terms, generating problems beyond the scope of intervention. He challenged his readers to "adjust our thinking to match the problem before us."[14]

*Solutionism*

While promoters of the techno-fix faithfully believe that technology is the best way to solve major problems, those engaged in solutionism are excited about ideas and inventions and look to put them to good use. The problem is they tend to develop solutions before they've defined the problem, basically putting the cart of the solution in front of the horse of the problem. The term "solutionism" has been around for a long time, but the person who popularized the term is the writer and researcher Evgeny Morozov, who studies the political and social implications of technology.

He writes that solutionism occurs when innovators presume rather than investigate the problems that they try to solve, reaching "for the answer before the questions have been fully asked."[15] As a result, problems remain poorly understood, insufficiently researched, or perhaps highly contested.

In Morozov's thinking, solutionism captures not only the flaws with underresearched and ill-defined problems; it includes situations when solutions drive the problem. In many ways solutionism thus originates in tech culture itself. Born of computing, the tech industry has revolutionized how we communicate, conduct business, and learn—all enabled by the exponential increases in the speed of microprocessors and bandwidth of communication devices. However, the proliferation, expansion, and enhancement of computing technologies has led to a situation such that these technologies could do little else besides make everyday transactions more efficient. Accordingly, many do-gooders, Morozov suggests, begin with an already conceived product or device and then go searching for a problem for which it can be put to use. With his focus on internet technology, he provides the example of apps that monitor such things as food intake and exercise activity. The developer of such an app has to sell a product that may not be obviously useful since many people self-monitor their food and exercise habits without the use of an app, or choose not to do so at all. So after the developer devises the solution (the app), they frame the problem as insufficient personal knowledge of one's behaviors.

For Morozov, problems come into existence because they are rendered "solvable," even "problems that hardly exist."[16] That, however, is not necessarily the case with food and agriculture. Sometimes the problems *are* serious—from depleted soils to lack of access to healthy food—but the solution is off the mark.

OpenAg's food computer provides a good example of solutionism. The inspiration of the Fukushima nuclear disaster notwithstanding, it sure looks like its inventors got excited about making a computer that could collect data on climatic conditions and then retroactively dreamed up a problem to which the computer could be the solution: a way to provide "hyperlocal" access to Tuscan tomatoes.

A related and perhaps more fitting idea is what political scientist Maarten Hajer calls "problem closure," referring to the tendency to define problems in relation to socially acceptable solutions.[17] Hajer argues that the nature and causes of many environmental problems in particular are contested, so actors reduce and simplify the problem in relation to what they can offer. As a result, they necessarily put aside certain aspects of the problem, including those that might lead to different and perhaps more difficult ways to address it. Here again, climate change provides a useful example. Climate change is obviously a highly complex, multicausal, overwhelmingly consequential problem. But if you render the problem as insufficient knowledge of your $CO_2$ footprint, then an app that tracks that footprint becomes a solution. Need I say that the proliferation of such apps hasn't made a dent in climate change?

Or consider another problem that is distressingly all too common in the United States: police killings of Black people. Attributing that phenomenon to "racial capitalism" does not lend itself to solutions. But if you (dramatically) circumscribe that problem to mean insufficient public visibility of these killings, then you can create a technological solution, say, video cameras worn on police officers' bodies, allegedly to create more accountability. As plainly evident from all-too-frequent news, body-worn cameras have done little to curb this kind of police violence. Although I suppose you could argue that there would be even more police violence without them, body cams are clearly not equal to the task. Even in more seemingly innocuous situations, when innovators back into problems based on their already conceived or socially acceptable solutions, they effectively leave by the wayside any serious intellectual engagement with the problem and therefore the possibility of more suitable approaches.

*The Will to Improve*

The third pillar of the problem with solutions inheres in the will to improve. Those acting with the will to improve have excellent intentions and often act from a place of deep kindness and even recognition of their own privileges in access to education or other resources. The problem is that they often act on behalf of others, without the latter's request, advice, buy-in, or even knowledge. The will to improve can manifest as a sense that you know best what others need, sometimes even if they tell you otherwise or ignore you altogether. This is a tough critique to make, since you're probably reading this (and I'm writing this) because we genuinely want to make the world more livable for other humans and other creatures who inhabit the planet. And we presumably want to sustain this livability for generations to come. The will to improve has some roots in the American tradition of charitable action. Supporting others is a good thing, but it has often been accompanied by efforts to reform the ways of vagrants, immigrants, and others who appear not to share white middle-class so-called "American values."[18]

The will to improve is most associated with the project of "development." In fact, the scholar who coined the term, the anthropologist Tania Murray Li, has conducted long-term research on development initiatives in Indonesia. Development as a project emerged out of the postcolonial era. Colonialism involved the extraction of wealth and resources from regions in the Americas, Asia, Africa, and oceanic islands that enabled capitalist development of the primarily European colonizers. While colonialism took different forms—not all was "settler colonialism" in which newcomers displaced Indigenous people—it nearly always left colonial subjects and Indigenous people worse off.[19] Following decolonization, most of which occurred by the end of World War II, the idea of development was born. Whether born of guilt, genuine concern, or the imperative to expand capitalism—all have been argued—development was an effort to improve standards of living in the former colonies. Since industrial capitalism and agricultural modernization had by that time proved to have considerably improved standards of living for the colonial powers, they generally provided the model of improvement. This was a different will to improve than that of earlier periods, when European missionaries sought to convert Indigenous people to Christianity.

Li's notion of the will to improve centers on the action of what she calls "trustees." Referring to politicians, bureaucrats, aid donors, specialists in agriculture, hygiene, and conservation as well as nongovernment organizations of various kinds, Li notes that "trustees" are benevolent in their intentions to improve landscapes and human well-being. Yet they tend to frame problems in terms amenable to their own expertise and values.[20] She characterizes this as a two-step process: problematization followed by "rendering technical." "Problematization" involves naming and characterizing the deficiencies that need rectification—for example, malnutrition, lack of income, or poor agricultural productivity. Problematization, Li notes, anticipates the kind of interventions that experts have to offer, similar to solutionism but with stronger intentions of doing good. Here you can see how founder Lisa Curtis of the Kuli Kuli operation had a strong will to improve.

The second step in the will to improve, intimately tied to the first, is "rendering technical." Closely aligned with the techno-fix, this refers to a set of processes of bounding the field, assembling information, and devising techniques to make the problem actionable, with the tools experts have at their disposal. For example, "rendering technical" might mean introducing high-yielding crops or, to return to the Kuli Kuli case, setting up an export scheme for moringa. Through these processes, Li argues, trustees attempt to improve the capacities of the poor (e.g., to become exporters of moringa) rather than address the practices "through which one group impoverishes another" (such as household gender inequality).[21] Like other scholars, Li notes that addressing the latter would take political contestation, which is why she writes that "questions that are rendered technical are simultaneously rendered nonpolitical."[22] The subtle distinction of rendering technical with the techno-fix is that whereas circumventing politics is the *objective* of the techno-fix, it is more the *effect* of rendering technical. This is, if you render a problem amenable to what experts can provide, their expertise rather than politics effectively drives the problem. For example, if an agronomist has the tools to make your crops more productive, they will define your problem as one of unproductive crops and you will no longer look to inequitable access to resources such as good land or credit as causes of your poverty. This facet of the will to improve is persistent and ubiquitous.

The will to improve pervades the growing field of global health. "Global health" refers to sincere efforts to address health disparities throughout the world. Although global health problems often stem from poverty or environmental injustice, global health solutions tend to center on diagnosis of particular diseases, containment efforts, or prevention through health education. The solutions to global health inequity are driven by what experts are capable of addressing. Given the origins of the will to improve, it is not surprising that its interventions have been described as neocolonial. Since colonialism was so deeply intertwined with racialization, referring to the impulse for colonizers to ascribe racial difference to justify colonial exploits, so too does the will to improve have a racial dimension.[23]

Writing in *The Atlantic*, author Teju Cole names it the "White Savior Industrial Complex." The term itself is a modern extension of the White Man's Burden, an ideology first named in a Rudyard Kipling poem, which rationalizes colonial exploits in the name of modernizing and improving the lives of others. Those burdened with the White Savior Complex are attuned to the damage of colonialism but can't rid themselves of belief in their own benevolence. Indeed, they find emotional satisfaction in believing they do make a difference. Nevertheless, as Cole retorts, "there is much more to doing good work than 'making a difference.' There is the principle of 'First do no harm.' There is the idea that those who are being helped ought to be consulted over the matters that concern them." Cole continues: "If we are going to interfere in the lives of others, a little due diligence is a minimum requirement."[24] Yet both the will to improve and the White Savior Complex foreclose genuine engagement with those who will be most affected by a solution.

In short, the problem with solutions is that they do one or more of the following: apply technology toward complex problems that demand political solutions; underresearch or neglect the underlying problem to sidestep what might best address it; and prioritize the ideas, values, needs, satisfaction, and expertise of those delivering the solution rather than the intended beneficiaries.

## *Solutions to Solutions?*

In recognition of at least some of these issues, some innovators and their critics have advanced alternative approaches—solutions to the problems with solutions, if you will. One prominent example is responsible innovation, which promises more substantive public engagement with contemporary processes of innovation. More salient in Europe and Australasia, owing in part to the stiffer opposition some technologies have received in those places, the principles of responsible innovation include anticipation, reflexivity, inclusion, and responsiveness.[25] This means attending in advance to potential consequences or public concerns, reflecting on the work as it is taking place, garnering the participation of those who might be affected in this reflective work, and responding to challenges. Responsible innovation thus requires ongoing dialogue between innovators and stakeholders as well as a serious commitment to make that meaningful. While an improvement, responsible innovation still assumes that technology is the answer. Indeed, it primarily provides a set of tools to mitigate potential objections to major technological developments such as geoengineering, effectively affording such developments legitimacy.[26]

Another approach is design thinking, purported to avoid many of the pitfalls of solution-making. Unlike responsible innovation, design thinking is routinely referenced in innovation sectors and often taught in tech-oriented university programs as well. With origins in mid-twentieth-century psychological studies on creativity, design thinking has since become a prescription for ways of approaching innovation that supposedly produces better solutions, even suitable for addressing difficult and multifaceted problems.[27] The modus operandi of design thinking is to ascertain what users want and then ideate and test solutions for feasibility and economic viability for these users, with these users. Specifically, design thinking prescribes an iterative process of empathizing, defining, ideating, prototyping, and testing.[28] Unlike solutionism, the process involves defining a problem in advance of formulating a solution; unlike the techno-fix, it is "human-centered," never presuming a technological fix is best; and unlike the will to improve, it makes a serious attempt to ascertain the needs of potential users.

Even at its face, however, design thinking considers its human targets in a very limited way, treating them as "users." Users connote those who might desire to take up a technology or program, not those who might be affected regardless of whether they were consulted. Moreover, design thinking explicitly avoids a fulsome problem analysis and purposefully omits from consideration solutions that require structural changes—an omission deemed a virtue. Instead, design thinking sees problems as resulting from a lack of creativity and innovation. Incapable of addressing problems that involve serious conflicts of interest—those that by nature require political solutions—design thinking aims for more frictionless processes. It sounds great, and who doesn't want to avoid conflict, but unfortunately few grand challenges lack conflicting interests.[29] Lee Vinsel, a steadfast and frank critic of design thinking (though decidedly not the field of design), writes that "the DTs" have permeated arenas outside of design, including efforts to make students changemakers. Noting the profuse use of Post-it notes as a key feature of design thinking, Vinsel finds it exceedingly vapid and cites many designers who don't take it seriously either. For him, design thinking is another version of the will to improve, "an elitist, Great White Hope vision of change that literally asks students to imagine themselves entering a space to solve other people's problems." Perhaps Vinsel's strongest concern is that design thinking gives students an unrealistic idea of creating positive change.[30]

Both responsible innovation and design thinking, in other words, attempt to improve on some of the problem with solutions but still effectively re-create them, by favoring technological approaches, bounding the field of problem analysis, and addressing those affected in a limited way. If they are not the antidote to the problem with solutions, what is?

*Response*

The economist Gilles Paquet offers a quote that has stuck with me. He writes: "Solutionism [interprets] issues as puzzles to which there is a solution, rather than problems to which there may be a response."[31] It is an idea that suggests another approach than solutions. Response, as I ponder it, starts from a different place than solutions: not what can I do, but what does this situation need? Assessing the situation in turn calls for an accounting of problems in their many dimensions rather than a rush to contain them. Response thus leaves room for recognition that many problems worth addressing stem from structural inequality and deeply ingrained sociological problems, including colonialism, capitalism, racism, sexism, and so forth. Response thus acknowledges that efforts to address the problem could likely help some and harm others, making implausible the circumvention of politics. Response might include technological innovation, but that is not the starting place. And because it does not preordain a technological fix, or shy away from politics, response opens up whole new worlds of action that solutions foreclose.

In fact, unlike design thinking, we can think of response as a true inverse of the problem with solutions: taking the time to consider the problem in its entirety and for whom it is a problem before acting (contra solutionism); recognizing the root social causes and vested interests in the problem and considering how those could be changed (contra the techno-fix); and asking that changemakers account for their own positionality vis-à-vis the problem and those it harms in order to consider who should lead the effort and under what conditions (contra the will to improve). As I imagine it, however, response does not mean that everything, everywhere, needs to be done all at once. Response, instead, takes strategy, which includes determining how to accomplish an objective in the immediate future, from which you adjust and build as conditions change. Strategy is what allows a problem to remain big and amorphous but the path forward to be specific and doable. Nevertheless, it is solutions that have taken hold, in many sectors and fields, and much of this book explains how hacking is not going to cut it.

## Contextualizing the Imperative of Solutions

As discussed, the impulses behind solutions have longer and more dispersed origins, but recently it has become impossible for changemakers to see past solutions as the way to address worldly problems. The privileging of solutions and only solutions didn't come out of the blue. To the contrary, the imperative of solutions must be contextualized in what some call "market world" and we academics call neoliberalism, referring to both an approach to governing the economy and a cultural zeitgeist.[32]

Neoliberalism as an approach to government originated with the Chicago school of economics, headed by Milton Friedman and Friedrich Von Hayek. Both economists believed that markets were far superior to states in allocating goods and services and thought that government should be shrunk to eliminate its interference in the ways of the market. As Friedman famously said, "Many people want the government to protect the consumer. A much more urgent problem is to protect the consumer from the government." In making their arguments, the neoliberals were responding to the huge growth of the state in regulating businesses and expanding social services, trends launched in the United States with the New Deal in the 1930s and expanded with the launch of the Great Society programs intended to redress poverty in the 1960s. They paid little mind to the fact that a primary cause of the US state's expansion owed to massive military expenditures in fighting both World War II and the amorphous Cold War, the latter inclusive of interventions in Korea and Vietnam. The neoliberals also ignored that this state expansion was primarily responsible for the United States' so-called golden age of capitalism, which saw unprecedented economic growth, rising incomes, and a reduction in economic inequality.

These early proponents of neoliberal ideas weren't taken very seriously until a set of political economic conditions began to threaten the stability of this golden age, beginning in the late 1960s and well into the 1970s.[33] These conditions included a huge balance-of-payments problem. Several East Asian economies, which had access to cheaper labor, had outcompeted the United States in key manufacturing sectors, such as electronics and automobiles. Thanks to the strength of the dollar, US consumers were spending their money on imports, while US exports had slowed. As

US manufacturing businesses began to falter and see profits decline, a once well-situated white working class saw their jobs disappear. Aggravating the situation, inflation was rampant in the 1970s—a result of OPEC, a cartel of oil-producing countries that drove up the cost of oil, and a series of unprecedented sales of US wheat to the former Soviet Union, in efforts to thaw the Cold War. Yet, unlike normal times of inflation, the US economy was not growing, instead beset by so-called "stagflation." Among other things, inflation was particularly harmful to those on fixed incomes, including senior citizens. When all of these issues came to a head, the ideas of those earlier philosophers appeared to make sense. Perhaps "big government" was the problem and not the solution.

While Prime Minister Margaret Thatcher of the United Kingdom was the first to embrace neoliberal approaches to government, Ronald Reagan's presidential election in 1980 changed the course of US history. Reagan had campaigned on racial resentments, invoking the specter of a Black "welfare queen," as well as grievances around inflation and taxes. In the early years of his presidency he attacked the power of labor unions and the impositions of regulation, blaming them for the loss of jobs to other countries. Reagan's agenda of tax-cutting, labor union–breaking, regulation enervation, and welfare reduction was never fully realized, although he certainly made inroads. In fact, some of the largest tax cuts, which largely benefitted wealthy people, were delivered by the two Bush presidencies (George and G.W.) and under Donald Trump, while the largest cut in welfare programs was undertaken during the Democratic presidency of Bill Clinton.

These tax cuts in particular set the United States on a course of less support for key functions, so that government-funded institutions—from schools to parks—needed to increase fees or find private sources of support. More distressingly, cuts to social assistance for the poor led to unprecedented levels of inequality in income and wealth, and huge swathes of the population living in precarious economic conditions. Moreover, cuts to or nonenforcement of the modicum of environmental, health, and safety regulations that existed worsened both public health outcomes and environmental quality while emerging issues such as climate change saw essentially no substantial government action. These areas eventually became fodder for solutions.

As a policy approach, neoliberalism was nevertheless successful at restoring business profitability, and this was especially true for the tech sector.[34] The experience of losing out to foreign competition in consumer electronics (such as radios, televisions, and small appliances, not to mention automobiles) made the United States want to be well-situated for emerging frontiers of innovation. Arguments that excessive regulation had stifled business innovation won the day, and so as a matter of policy, the US government opted to give both the incipient biotechnology and consumer electronics industries virtual free rein. Receiving outsized approbation and all sorts of government-bestowed advantages, by the early 1990s the US tech sector came to be both the economic powerhouse and cultural phenomenon it is today, with Silicon Valley as the centerpiece.

The tech sector's stock-in-trade was of course solutions. This should not surprise. Problem-solving is at the core of innovation. Furthermore, developments in computing, logistics, and communication can provide all manner of business solutions, understood in that techie way: higher speed, better coordination, ease of use, and so forth. Yet by the early 2000s techies began to tackle problems beyond their historical purview. Impressed with their own can-do spirit, their faith in disruption, and called upon to engage higher-order issues, the tech sector began to take on social and environmental issues—issues that have historically been the province of social movements, civil society organizations, universities, and the public sector.

Neoliberalism had meanwhile changed how people think about the role of the public sector in relation to the private sector. For one, it altered how people view the capacity of government to act on behalf of the public interest. The constant drumbeat about the shortcomings of the state, coupled with the underfunding of services, contributed to their actual decline. A loss of faith in the state encouraged a turn to private individuals and entrepreneurs as progenitors of positive change. Many of those who had amassed enormous amounts of private wealth, thanks to unbridled profit-making in the tech sector and beyond, developed a sense that they need to "give back."[35] So did the many celebrities who became high-profile campaigners for human rights in acts of so-called "celanthropy." Much of this giving went to causes of their own choosing, underwriting solutions rather than fundamental social change.[36]

Yet even those without access to wealth started to believe that the private sector was the best avenue to bring improvement. This gave rise to the growing field of social enterprise, characterized by "triple bottom line" business models to achieve social, environmental, and financial returns. Guided by a will to improve, but also needing to make a profit, these so-called social entrepreneurs turned to delivering solutions, narrowly construed. This is in no small part because such businesses have to sell a product (or service) to generate revenue. Consider the many shampoos, lotions, cereals, stuffed animals, or consumer goods marketed around saving the rainforest. It's hard to fathom how buying products derived from rainforest species actually helps save the rainforest, even if a small percentage of profits go to rainforest conservation.[37]

For that matter, neoliberalism gave rise to the idea that entrepreneurialism itself could be a road out of poverty. Rather than being dependent on the state for housing or food assistance, poor people could improve themselves by developing businesses of their own, so it was imagined. Indeed, much of the new philanthropy was aimed at those who have been harmed by letting markets work, but under the premise that they had simply been left out, often emphasized providing them the tools of the market.[38] As such, many organizations shifted from directly supporting poor people to "empowering" them to be more self-sufficient. Consider microfinance, in which progressive, presumably altruistic financial institutions offer small loans to potential entrepreneurs, often women. And yet, after enjoying a heyday of praise for its creative approach to third-world poverty (and a Nobel Peace Prize awarded to the person who pioneered the idea), microfinance became subject to scorn. These small loans saddled borrowers with unmanageable debt, especially when their products found no markets. In India a spate of suicides was attributed to unpaid microfinance. Still touted by major figures in the field of development, the problem with microlending as a solution is that it problematized poverty in a very narrow way that was based on the assumption of discrimination in business lending.[39] Kuli Kuli directly inherited this neoliberal-influenced "pro-poor" approach that encourages those in precarious positions to become entrepreneurs and potentially exacerbate their precarity.

The imperative of solutions has permeated the nonprofit sector as well, even grassroots nonprofits. In the United States one reason such organizations develop solutions is that they must restrict their activities to educational or charitable programming to maintain their tax-exempt status. They therefore develop programs and activities that are relatively apolitical and try to show how these activities connect to a larger problem. Being beholden to fickle funders also lends itself to solutions. In their attempts to instill accountability, government grantors and philanthropic organizations often insist that such organizations produce "deliverables" to receive and maintain funding. Deliverables tend to be short term, easily quantifiable outcomes toward specific objectives. Accordingly, nonprofit organizations may prioritize activities that translate as solutions rather than those that organize for the long haul to contribute to a substantive goal. An organization that wants to address homelessness, for example, might develop a database on housing availability or train a certain number of people on how to find housing. Even if organizational leaders know the root cause of homelessness to be poverty, funding requirements prevent them from working in the political realm to pass antipoverty legislation. Government regulatory agencies have come to prioritize solutions, too, favoring pro-industry projects over regulatory restrictions.[40]

More generally, neoliberalism encouraged a veneration of private sector economic activity such that all other institutions, including the university, became expected to serve it directly. In the past, universities invested in more research that may not have had obvious applications at the time it was conducted. In addition, universities generally balanced applied work with liberal education models that train students in critical thinking and humanist concerns. With neoliberal influences, however, universities came under increasing pressure to prove the immediate relevance of research before it was even conducted. Moreover, with taxpayers less willing to support them, universities had to garner enough extramural funding to stay afloat. And so universities began to emphasize science, technology, engineering, and math (STEM) education, which they assumed would have more applicability in "the real world" than social science, arts, and humanities. The pressure of MIT's much-hyped OpenAg Initiative was in part tied to the power of such initiatives to attract external support at a time when universities increasingly compete for philanthropic and corporate funding.

In short, many outcomes of neoliberalism converged to elevate solution-making as the preferred mode of action to redress the fallout of a political economy that decreasingly protected the population from economic precarity and illness and the environment from harm. And yet it also put constraints on how solutions could be pursued, giving pride of place to the most dynamic parts of the economy and those who were succeeding by it—in other words, the tech sector and those who were willing to operate under the same terms—and with a sensibility that pursuing these narrowly constrained solutions was the right thing. As the Silicon Valley-style tech sector became ever more culturally salient, it inspired an entire belief system about how to approach change, one that permeated other institutions—among them the universities that not only were victims of the changing political economy but were now responsible for creating changemakers of the future. The proliferation of solution-oriented programs on university campuses is the surest sign of the increasing influence of the tech sector on academia. You may well be reading this book because you're enrolled in such a program.

And yet, as this book shows, Silicon Valley-style solution-making is at best ineffective and at worst harmful. One of the ways it is harmful is that it has penetrated the culture so much that we see no other options. My goal is to get you to resist the Silicon Valley problem-solving sensibility and develop a capacity for and interest in response instead. Indeed, it is crucial that you do so given the current moment in history. On that note, it is increasingly clear that neoliberalism is no longer exactly the guiding principle of government, as left-leaning parties are revisiting, albeit haltingly, New Deal-style measures (as evident in the COVID relief and infrastructure bills), and right-leaning parties are flirting with fascism (as evident in measures to ban books from school and universities and in policies restricting the rights of women, immigrants, and transgender people, and enervating democratic governance institutions). In this context, especially, solutions are completely ill-suited to the task.

All that said, the problems with solutions may be nowhere more evident than in the massive growth of efforts to transform food production and distribution, best indicated by Silicon Valley's recent investments into this domain and the many university programs springing up to hack the future of food. My most recent research project—a collaboration of geographers, sociologists, and cultural historians—has looked closely at the Silicon Valley tech sector's foray into food and agriculture. Our research has asked how the tech sector addresses the challenges of food and agriculture with a specific focus on how it defines problems and solutions. This has involved using publicly available sources to collect basic data on hundreds of companies that identify with the agrifood tech sector, participant observation at nearly one hundred events, as well as interviews with over ninety start-up executives, investors, and industry consultants.

To cut to the chase, we find that despite grand ambitions, many of the solutions the sector brings are underwhelming, unnecessary, or untoward. Differing from prior alternatives to industrial food and agriculture such as organic farming, which mostly had the technology right but adopted the wrong approach for proliferating its uptake, much of this mismatch, our research findings show, stems from techie desire to use technology for impact while having minimal understanding of the dynamics of food and farming. Although that research animates much of this book, I have been researching efforts to transform food and agriculture for my entire academic career (decades, in other words), and that too informs the book, as do my experiences with UCSC as it increasingly embraces solutions in agriculture and food and beyond.

As such, most of the examples I bring to bear throughout *The Problem with Solutions* center on food and agriculture. However, the lessons within apply to spheres beyond food and agriculture and encompass the array of social and environmental crises areas now subject to an array of solutions. Some of these solutions will have little to no effect and some will undoubtedly change things, but that doesn't mean they will make things better. I therefore begin in chapter 1 with a deep dive into how Silicon Valley came to anoint itself the maker of solutions.

# 1   *Silicon Valley and the Urge to Make the World a Better Place*

With pointed satire, the pilot episode of HBO's hit comedy series *Silicon Valley* depicts the basic trappings of Silicon Valley start-up culture. A nerdy, awkward programmer named Richard Hendricks works for Hooli, a large tech corporation. Obviously modeled after Google, Hooli is led by CEO Gavin Belson, an Elon Musk sort of figure, full of ego and outlandish visions. As a side gig, Richard works on an app dubbed Pied Piper, which tracks songs to help musicians avoid copyright infringement. He lives in a suburban Palo Alto home that doubles as an incubator for aspiring entrepreneurs. The head of the incubator, the weed-smoking Erlich Bachman, once founded and sold a tech company. On that basis he mentors others in exchange for rent of the workspace. Other house denizens—all men, all white except for one Pakistani—possess varying levels of skill in programming and delivery of sardonic quips.

Eager to garner interest in his app, Richard pitches it to two "brogrammers" at Hooli as well as an outside venture capitalist. All dismiss it at first. But when the brogrammers actually take a look, they are impressed with the efficiency of the app's compression algorithm. This information makes its way to CEO Belson, who offers to buy rights to the application at $10 million. While meeting with Gavin, Richard receives a call from the venture capitalist, who offers $200,000 for 5 percent of Pied Piper, which would leave Richard in charge. Learning that Belson wants the algorithm under his control only to squelch it, Richard takes the venture capital offer and brings on others in the house to work on Pied Piper. A start-up is born.

While this first episode gives a snapshot of the Silicon Valley "ecosystem," another episode from the first season draws a finer point. This one portrays a pitch event, a format invented by Silicon Valley to connect wannabe entrepreneurs with potential venture capital funders. In direct imitation of Silicon Valley lingo and staging, the event is called TechCrunch Disrupt SF. As is typical, a series of white and Asian men take the stage and make short pitches about their products, while judges watch and prepare to dole out advice. Pitching products with technologies such as "paxos algorithms for consensus protocols," "crash analysis platforms," "location-based mobile news aggregations," and "software-defying data centers," virtually every pitcher punctuates the pitch with a statement about "making the world a better place." The scene climaxes when someone pitches a microwave technology that can heat a person's skin to save the energy of heating an entire room. At that moment the judges interrupt to point out that the technology sounds very dangerous and no one will ever buy it. Hilarity aside, the scene depicts something essential about contemporary Silicon Valley: an ever more evident imperative to improve the world. No longer enough to stand by Google's early—and now laughable—motto of "Don't be evil," companies must now make the world a better place. Or as Facebook cum Meta's Mark Zuckerberg once proclaimed, his primary goal was not to make money . . . but to help solve the world's problems.[1]

Until very recently, Silicon Valley's idea of solving the world's problems has meant facilitating and smoothing ways for the already privileged people to be in the world—for example, ordering food with an app rather than picking up the phone. And yet, moving in the direction of revolutionizing ways of living has brought the tech world ever closer to wanting to address the planet's profound ecological and social crises. How did Silicon Valley arrive at the goal of making the world a better place and how does the Silicon Valley ecosystem—a term used by the tech sector to describe its network of relationships—shape these aspirations? This chapter examines Silicon Valley's cultural and political economic history to explain its unique but highly influential approach to "making the world a better place," an approach that has become a model for solution-making.

## Where Is Silicon Valley?

Or better put, *what* is Silicon Valley? Geographically it sits on the southern end of San Francisco Bay, centered in towns like San Jose, Cupertino, Mountain View, Santa Clara, Menlo Park, and Palo Alto, the home of Stanford University. Its "valley-ness" stems from the fact that low mountains (by California standards) form the better part of the region's perimeter. Once planted in stone fruit orchards and dubbed the Valley of Heart's Delight, it obtained the moniker of Silicon Valley much later, when the semiconductor industry took hold. The Pruneyard Shopping Center is one of the few remaining indicators of the region's agrarian past. In sharp contrast to the high-rises and dense urban neighborhoods of San Francisco, it maintains the built environment of a post–World War II suburb, with sprawling housing developments and strip malls.

The tech industry has layered that landscape with office parks and "campuses," replete with low-rise buildings, parking lots, and somewhat sterile green spaces to accommodate the thousands of tech workers that pour into the valley every workday, pandemics permitting.

But the tech sector reaches far beyond this valley. The entire San Francisco Bay Area has become a magnet for tech, with businesses locating throughout: in San Francisco, Oakland, Emeryville, Berkeley, and other smaller cities around the bay and farther south into Santa Cruz. Moreover, Silicon Valley, while preeminent, is hardly the sole center of tech innovation in the United States. The Boston area actually had a jump start on Silicon Valley in high-tech industries, aided by its close relations with MIT and Harvard and, in fact, Boston hosts several organizations that have welcomed food tech culture especially. Seattle became a tech hub with Microsoft's founding long ago. Recently a number of other US cities with top universities have attracted technology start-ups, including Austin, Texas, and Columbus, Ohio. That these cities have become tech-centric owes much to the close-knit relationships with the universities and the synergies they bring; the universities train the tech workers of the future, while the tech sector funds tech-oriented university programs. Nor is the United States the only tech center worldwide. Innovation hubs are scattered throughout Europe and Asia; those in Norway and the Netherlands are on the frontier of food innovation.

Still, Silicon Valley the place forged a new way of organizing innovation and created an unparalleled culture of hype and hubris, making it much more than a geographic place defined (or ill-defined) by a particular physical geography. It is, as historian Margaret O'Mara puts it, a "global network, a business sensibility, a cultural shorthand, a political hack."² I thus refer to Silicon Valley as both a literal geographic place and a synecdoche—a stand-in for a larger phenomenon.

## A Short History

Silicon Valley's innovation culture did not begin with the iPhone nor by entrepreneurial start-ups, touting the latest and greatest way to solve some problem you didn't know existed. No, its physical transformation from plum fruit orchards to office parks began with a culture different enough from the current one to defy images of Silicon Valley as a hype-driven, go-it-alone, impact-focused hotbed of innovation. It began as a product of the war industry, once a linchpin of the California economy, with the lion's share of funding provided by the federal government in the form of military contracts.

### *Born of War*

The standard origin story begins with Frederick Terman, an electrical engineering professor at Stanford University. He encouraged two of his graduate students, David Packard and William Hewlett, to commercialize an audio oscillator that Hewlett had designed for his master's project. Terman lent the two some of his own money and arranged a loan from a nearby bank in Palo Alto. Out of a garage, Packard and Hewett founded the Hewlett-Packard Company in 1937, the very first electronic company on the West Coast. HP, as it was eventually called, first sold these devices to the Disney Company. Yet the company came of age during World War II, when it became a military contractor for the US government, supplying electronic measuring devices and receivers to detect and analyze enemy radar signals. As a result, the company's sales increased twentyfold between 1941 and 1945. Several other companies, producing such items as vacuum tubes and wireless transmitters, also with the help of Stanford University and federal government military contracts, were born in the valley during this same period. With Harvard and MIT having been the original beneficiaries of military contracts, these arrangements began to shift the center of gravity for the tech industry westward.[3]

Following the war, Terman's efforts solidified this shift. Terman saw huge potential in developing the valley as a base of the technology industry—and bolstering Stanford as a technology-oriented institution. Stanford held some key advantages over its eastern counterparts, not least of which were its massive real-estate holdings and the cheaper real estate in the surrounding area. (What a time it was!) Appointed dean of engineering, Terman forged entirely new roles for the university in relation to industry. Following the success of HP, he encouraged his students to move fluidly between industry and the university, offering them teaching positions in addition to support for their businesses. This dynamic readied the valley for the next infusion of funding. The Cold War between the United States and the USSR not only had made high-tech research a national priority; it generated another round of massive military expenditures, providing unseen opportunities for tech companies to develop and manufacture the necessary equipment.[4]

This early history of Silicon Valley, nurtured with government support, bears acknowledgment in the context of contemporary, neoliberal-inflected imaginations that "freeing" industry from the state will bring forth economic development. Indeed California's "second gold rush," a period of immense economic expansion, owes a great deal to federal investments in the war industry and the wealth it brought.[5] For that matter, even the first gold rush, springboard of California's first wave of economic development, wasn't exactly about freeing up entrepreneurial impulses; it rested instead on the discovery of an extremely valuable mineral in the public domain. Like the first gold rush, though, this second one created the conditions of extraordinary regional economic development. One of the core technologies developed during the Cold War was the semiconductor, a critical component of computer chips made with silicon. Semiconductor innovation was a mainstay of companies such as Intel and National Semiconductor based in the valley, giving rise to its name. As military and aerospace demand for semiconductors dwindled, a new set of enterprises began to take up the slack, centered on consumer-oriented technologies.

*Countercultural Computing*

As legend goes, the personal computing revolution was highly influenced by the 1960s counterculture and New Left, of which San Francisco was a hotbed. Cold War computing had both inspired rebellion—after all, the New Left's origins were in no small part born of opposition to the Vietnam War—and imitation, as the military-industrial research world had also given rise to a "freewheeling" and collaborative style of work.[6] Carving a niche between bohemian San Francisco and the emerging technology hub of Silicon Valley, and rejecting the cultural strictures of the 1950s and early 1960s, a cadre of journalists and other influencers championed networked computing as a tool for personal liberation, a way to build alternative communities and utopias.[7]

Loosely connected to that ethos, the Homebrew Computer Club first met in 1975 in a Palo Alto garage, bringing together nerds and alleged countercultural types. Some were students at Stanford, while others were dropouts; some had jobs at other companies, while some were working on their own. As the mythology goes, their cultural openness and willingness to take risks spurred norm-bending innovation. Among them were Steve Wozniak and Steve Jobs, who had the technical chops to develop a microcomputer for personal use. They launched the Apple Computer Company in 1976, with the creation of the first Apple computer. Apple computers' inventive form and function, along with their cute logos, gave them a sort of cultural currency that helped launch entire new directions in consumer electronics.[8] The Apple computer, along with the IBM PC, totally revolutionized computing, making owning and working a computer a near necessity for the professional world and eventually for everyday life.

Following those heady days, innovation began to occur at breakneck speed, creating a culture of change and excitement around technology itself. With a culture that valued risk-taking, inventors were not afraid to fail, and those around them did not shame them for doing so, making the world safe for ambitious entrepreneurs anxious to start up their own venture. The ensuing start-up culture valued technological achievement much more than economic success. As put by Anna Saxenian, who studied the Silicon Valley innovation sector in relation to Boston's, put it: "The elegantly designed chip, the breakthrough manufacturing process, or the ingenious application was admired as much as the trappings of wealth."[9] Even though many of these entrepreneurs went on to make millions, sometimes billions, of dollars, they continued to upgrade, transform, and invent.[10]

By the late 1980s Silicon Valley's ventures into consumer technologies had paid off, buoyed by the success of Apple and the many supporting companies making components and software. Personal computing promised enormous markets, as did computer gaming, computational devices, portable music systems, and so forth. Yet more innovation to increase the ease and speed of use followed, with smaller computers, larger-capacity hard drives, more memory, faster processing time, and ever more user-friendly software. Speed and efficiency of use were themselves revolutionary.[11] Whether such improvements were necessary or simply possible was a question rarely asked as technological innovation for its own sake increasingly infused Silicon Valley culture, paving the way for solutionism to take hold.

## A Different Kind of Government Support

By the 1980s Silicon Valley was delivering on its promise to bring postindustrial economic expansion to the United States, filling the gap left behind by the decline in manufacturing.[12] This was due in large part to the deliberate strategy of the US state, guided by neoliberal notions, to create the utmost in business-friendly environments. Policymakers, in fact, opted for a path of least resistance in the regulation of new technologies, to incentivize the entry of domestic firms. For example, a 1980 legal ruling in a famous case called *Diamond v. Chakrabarty* allowed the patenting of living organisms. The importance of patents for innovation is that they prevent the copy of technologies, meaning the company that invented the technology, product, or in this case, living organisms can sell their product dearly without fear of competition for the duration of the patent. Patenting and hands-off regulation were extra important for the biotechnology industry because of its slower development cycles. These liberal policies helped birth "Biotech Bay" alongside Silicon Valley, beginning with the debut of Genentech.

A minimalist approach to regulation also boosted opportunities in computing and electronics, particularly with the arrival of the world-changing internet. As it happens, the internet was another product of the US Defense Department, until the National Science Foundation took it over for academic use. Its rather late-coming commercialization in the 1990s precipitated whole new waves of innovation in e-commerce and social media.[13] Purposefully not put under centralized control so as not to curb innovation, the internet's emergent business model of monetizing personal data and propensity for security breaches owes to the lax regulatory oversight of the US government. An additional gift of the federal government to the tech sector was the increase in federal grants to universities, which provided both training and basic research that could apply in the private sector.

*Venture Capital Takes the Stage*

Another aspect of the more neoliberal-inflected political economy bolstered Silicon Valley's culture of solution-making: the windfalls of lower taxes and little regulation had created vast pools of wealth in search of profitable places to invest. This paved the way for the fluorescence of a new kind of funder to become the mainstay of Silicon Valley early-stage financing: venture capital.[14] Largely replacing military contracts, venture capital didn't exactly arise independently of the federal government. To the contrary, friendly taxation and financing programs, along with a hands-off regulatory approach, cleared the way for venture capital's rise.[15] Traditional banks had been highly reluctant to invest in those who worked outside an academic lab or didn't even yet have a product on the market (so-called vaporware, "to describe imaginary high-tech products announced with fanfare before they actually existed"[16]). The culture of accepted failure didn't fly, in other words, with traditional investors. Start-ups needed investors who were willing to take big risks in exchange for big rewards, to bet on unproven people, technologies, and products. Venture capital was the perfect match: venture capitalists invest in many companies knowing that most will fail, but they will gain massive payoffs if just a few succeed, of up to "two thousand percent," according to one venture capitalist.[17]

The Small Business Investment Act of 1958 had laid the foundations for venture capital involvement, providing a set of tax breaks and federal loan guarantees for investment in small businesses. This act sanctioned the creation of the Small Business Investment Company (SBIC). Through these vehicles the federal government would guarantee loans three times the amounts an SBIC would raise. By 1959 five hundred such companies were born, although many eventually took up a limited partnership model.[18] Following a 1978 change in the tax law reducing both corporate tax rates and capital gains taxes, venture capital investments began to flood the sector.[19] Meanwhile, the federal government's Small Business Innovation Research (SBIR) program effectively acted like private venture capital.[20] Begun in 1982, the SBIR program required that federal agencies with large research budgets devote 2.5 percent of their research and development (R&D) budgets to supporting spin-off firms.

Venture capital continued to grow in importance in the 1980s, thanks to its willingness to fund unproven high-tech ventures.

Recognizing that universities spur the kind of research that might lead to profitable investment, venture capital firms tended to congregate around research centers and universities, adding to the regional centralization of Silicon Valley. By 2000 there were more than ten times as many venture capital firms engaged in active investing as there had been in the mid-1980s, with the vast majority of financing going to firms based in the Bay Area.[21] As you will see, the specific demands of venture capital on start-ups encouraged the latter to frame their offerings as solutions.

## *The March of Disruption*

In the 1990s growth of the internet and e-commerce, as well as the invention of smartphones, brought new opportunities in software development and platforms. Cloud computing meanwhile lowered the cost of entry for new start-ups. To develop a platform or smartphone app, all you really needed was technical know-how, a laptop, and cash to rent server space. Start-ups could skip the Silicon Valley office parks and set up shop in shared work spaces, suburban houses (as depicted in the HBO series), or even their own apartments.[22] All of this led to the proliferation of platforms, apps, and social media sites, making innovation more about networking than silicon. By some estimates, more than two-thirds of venture capital funding went to internet companies in the late 1990s.[23]

Many e-commerce companies didn't work out. A dot-com bubble led to a dot-com bust when many new e-commerce sites failed to turn profits, causing their stock to freefall. Sometimes, however, these platforms and apps provided services that changed the way we live in the world, most prominently demonstrated by the likes of Amazon, Uber, and Airbnb. Amazon all but erased brick-and-mortar bookstores, Uber virtually demolished the taxicab industry, while Airbnb put a serious dent into the fortunes of the hotel business. Such start-ups also ushered in the demise or at least diminution of phone booths, video stores, paper maps, cash payments, record albums/CDs, cameras, and more, leading a prominent venture capitalist to write in 2011 that "software is eating the world."[24]

To make sense of these profound and sometimes harmful changes, aided and abetted by the ready availability of venture finance, many newfound companies turned to the notion of "disruptive innovation." The framework of disruptive innovation was first championed by Harvard business professor Clayton Christensen. Grappling with how otherwise good businesses fail, Christensen had suggested that extraordinary entrepreneurs can and *should* rejuvenate industries that are technologically backward and resistant to change with both new business models and products.[25] Christensen later lamented that the theory had been misapplied to describe "any situation in which an industry is shaken up and previously successful incumbents stumble."[26] No matter his more precise meaning, the term was already widely in circulation, with many entrepreneurs claiming disruption as a selling point for their innovations. As it happens, not all new ventures were situated to be *that* disruptive, but new entrepreneurs learned to *narrate* that their technologies would be, as if disruption itself gave them moral grounding.

## Toward Really Making the World a Better Place

By the turn of the twenty-first century, the ethos of Silicon Valley had morphed into making the world a better place. It is not as though certain tech giants didn't already believe that they were bringing improvement. Mark Zuckerberg had long held that his social network would promote empathy and connect people, just as Sergey Brin and Larry Page had insisted that Google not be evil. Nevertheless, it had become nearly de rigueur that tech companies claim they were improving the world.

Silicon Valley's moralizing reflex was in part inherited from the earlier utopian and countercultural movements that once centered in the region. These not only imagined and attempted to model better ways of living; some former communards, disenchanted with communes, turned to computing as salvation.[27] More amorphous aspects of Bay Area culture rubbed off on Silicon Valley as well: a laid-back attitude in relation to the east, a sense of intense optimism, and a rough belief that betterment was always possible. What came to be referenced as "the Californian ideology"—a "bizarre fusion of the cultural bohemianism of San Francisco and the high tech industries of California"—borrowed ideas from both the political left and right, combining individual antiauthoritarianism with "a profound faith in the emancipatory potential of the new information technologies."[28]

The tech sector's moralizing impulses also stemmed from the proliferation of innovation in search of need or meaning, an attempt to reconcile moral worth with profit-making.[29] This, however, often manifested as a conflation of what the tech sector could offer with progress.[30] Consider what "making the world a better place" actually had come to mean. For most of Silicon Valley's history, it had decidedly *not* meant using technology for environmental betterment or social equity. Rather, "improvement" had connoted a reduction of the frictions of time and space in everyday transactions, as if the most challenging aspect of living is that things take too damn long. To be sure, what Silicon Valley had best delivered on is efficiency—in computing, communicating, logistics, and more. Uber, for example, may be a more efficient way to hail a ride, but it does not take more cars off the road to reduce carbon emissions. It may be technological progress, but it is not an environmental improvement.

Still, the turn toward making the world a better place was not just a matter of self-justification. Some in the Valley began to express genuine interest in social and environmental improvement, acknowledging that unbridled capitalism had created something of a mess and that they, perhaps following on the heels of the bit tech moguls cum philanthropists, had the creativity and resources to do something about it. Disillusioned by the excesses of the dot-com era and embracing a newfound will to improve, many companies shifted the focus from changing the world (aka "disruption") to saving the world. Perhaps the clearest indication was the sizable investment that Silicon Valley put into the cleantech sector (also called green tech), beginning in the early 2000s.[31] "Cleantech" broadly refers to the use of technology to solve environmental problems; the primary target of this unapologetic use of the techno-fix has been the energy sector, with aspirations to replace carbon-emitting fossil fuels with renewable sources to address climate change.

John Doerr, a quasi-famous venture capitalist at Kleiner Perkins in Menlo Park, was one of cleantech's biggest champions. Having seen Al Gore's *An Inconvenient Truth*, Doerr invited Gore to be a partner in the firm. His thinking was that Silicon Valley needed to move beyond consumer software and take on global "grand challenges." In a passionately delivered TED (technology, entertainment, and design) Talk in 2007, Doerr proclaimed that investment in green technologies would be "the biggest economic opportunity of the 21st century," and, if successful, these technologies would make the "most important transformation of life on this planet."[32] Doerr was not acting alone. Many investors and entrepreneurs who had "ridden the internet bubble" began "pouring money and ideas into cleantech." This included the cofounder and former CEO of Sun Microsystems, as well as PayPal cofounder Elon Musk, who invested $96 million of his own money into his electric-car start-up Tesla Motors.[33]

The year 2007 turned out to be an important inflection point, marking the fluorescence and unprecedented legitimacy of green capitalism. While businesses with social conscience (think Ben and Jerry's) had been around for decades, the new green capitalists were thinking much bigger and welcomed the opportunity to work within capitalism to change capitalism.[34] They wanted to do more than donate a percentage of profits to a good cause or act as model employers. The 2007 advent of the "B" corporation, a business form that allows companies to address social and environmental benefits beyond its shareholders, is yet another indicator that having your impacts and profiting from them too was an idea whose time had come. That same year, a proliferation of a new kind of investor calling themselves "impact investors" made ever larger pools of funds available to this new breed of start-ups.[35] Building on earlier iterations such as socially responsible investing, in which institutional funds attempt to minimize the presence of bad corporate actors in their portfolios, impact investors began to directly finance businesses that aim actively to solve social or environmental challenges.[36]

Silicon Valley thus became a place where solutions are made. The kinds of solutions that were made, however, were a logical outcome of the political economy and cultural zeitgeist that so defines the region: an antiregulatory economic development engine focused on technology that has been made palatable by its association with California's countercultural past and socially liberal climate. This political economy and cultural zeitgeist were fertile ground for the impulses behind solution-making: a climate that would apply technology to address problems while steering clear of fundamental social change (the techno-fix), a love of invention to see where it leads (solutionism), and a desire to do good albeit in ways defined by innovators' desires and abilities (a will to improve).

And yet, when all is said and done, the Silicon Valley model has not proved to be particularly conducive to delivering on its newfound social and environmental commitments. A closer look at the Silicon Valley ecosystem can help us see why.

The Silicon Valley Ecosystem

As suggested earlier, Silicon Valley start-up culture evolved to foment risk-taking and creativity in innovation. At the same time, the ecosystem that developed to facilitate that creativity revolved around the achievement of what is called an "exit." This refers to a situation in which either a start-up, having proved a product or process viable, sells its technology to a well-established corporation or it takes the company public to be sold on a stock market. It is only through an exit that start-up founders have a chance of making the big bucks; it is only through an exit that venture capitalists and other funders get paid off for the risks they have taken in the early stages of the company. For their part, corporations who buy out promising start-ups effectively outsource their R&D. Rather than be saddled with the enormous expenditures of developing new product lines or manufacturing processes in-house, they look to venture capital to vet promising technologies.[37]

A consideration of some of the key players and how they facilitate the exit will begin to show why this ecosystem is not so favorable for making technologies that can deliver significant environmental or social improvement, even when intentions are sincere.

## Incubators and Accelerators

Many very small start-ups begin like the fictional Pied Piper, in an incubator. The incubator is an organizational form that proliferated with the dot-com bubble of the 1990s. Like the name invokes, incubators work to nurture infant start-ups before they are ready to survive on their own. Sometimes organized as nonprofits, incubators may provide as little as a coworking space and access to the internet and a file server. They may provide as much as mentorship from accomplished professionals or, in Richard's case, someone who had a successful exit. In return for the workspace and mentorship, the wannabe start-up pays rent, often for a long duration, until they find a product or process that might work. Their aspirations could well die in the incubator.

More promising companies may move on to an accelerator program, the first known of which was created in 2005. Unlike incubators, accelerators may be more selective in who they take on, in no small part because they often expect to take some sort of equity position if the company becomes successful—meaning they will own a share of it. Accelerators' programs are more intensive than incubators'. They aim to support new ventures by shaping their initial products, identifying potential markets, and securing resources, including capital and perhaps employees. They usually provide a small amount of seed capital as well. Most important, accelerators provide introductions to both peer ventures and mentors, who might be successful entrepreneurs, program graduates, venture capitalists, angel investors, or even corporate executives. Many accelerators culminate with a "demo day," such as a pitch event to an audience of qualified investors all for the purpose of raising additional funds and perhaps eventually a successful exit.[38]

*Venture Capital*

Venture capital greases the wheels of this system, providing early-stage funding for the company to refine their product and business plan to prepare for the exit. But venture capital is selective in who they fund, even knowing that many they fund will still not succeed. Mainly, they want to be convinced that their overall portfolio of companies will be profitable. Generally venture capital looks for three things in a start-up, although they may emphasize one: people, technology, or market size.

In emphasizing people, they often look for a spunky entrepreneur with a good story (witness Kuli Kuli). In emphasizing technology, they often look for a technology that will be groundbreaking and not easily imitated. Here, they are particularly attracted to technologies that are patentable or are able to obtain other intellectual property rights protections. In focusing on markets, they want a technology with the potential to reach vast numbers of potential users, whether other businesses or consumers. Once they fund a start-up, they tend to take a hands-on approach, as they have a direct interest in ensuring success. So they will work with the company on their business plan and help the company further its networks. Their involvement may even signal to incumbents that the start-up is worth acquiring.[39]

## Pitch Culture

The dynamics among these actors are on full display at one particular moment in this process: the iconic pitch event—the phenomenon described in this book's preface and parodied in *Silicon Valley* at TechCrunch Disrupt. These pitch events are more than interactions between start-ups and venture capital, facilitated by accelerators; they are public performances of a communication style also born in Silicon Valley. This upbeat style is not incidental to Silicon Valley culture but its greatest manifestation, demonstrating the centrality of hype, positivity, and ambition in promising better worlds. The style was popularized by Steve Jobs, who when offering up a new product would stage an event to excite the public. Dressed in his signature black turtleneck, Jobs would enter the dramatically lit stage, look piercingly into the audience, take the microphone, and tell a story about the future. He would then introduce the gadget that would change life as we know it. The dramatic performance was supposed to conjure up confidence that whatever was on hand *was* the future and could not be missed out on. Others imitated this performative style, notably Elizabeth Holmes of Theranos fame. It also provides some of the most comedic moments for the Gavin Belson figure on *Silicon Valley*.

This performance style further proliferated in the now widespread TED Talks. These highly rehearsed and staged short lectures, seventeen minutes to be exact, attempt to be both inspirational and pedagogical in their presentation of big ideas and big solutions. They reflect a Silicon Valley culture of "thought leadership" in which easy, actionable ideas are favored over reasoned critique and traditional expertise.[40] For some they epitomize an anti-intellectual, simplistic approach to problem-solving. One wry professor of visual arts from the University of California–San Diego used his own TEDx Talk to exclaim that "TED actually stands for: middlebrow megachurch infotainment."[41] Nevertheless this kind of performance has become an expectation at the ubiquitous pitch events in the Valley and its environs, meant to exude the confidence, determination, and, yes, world-changing aspirations of the pitcher to attract the capital they need to make their world-changing technology a reality.[42]

The pitches themselves, ranging from four to ten minutes, are remarkably formulaic and rehearsed. Typically the entrepreneur begins by articulating a problem that must be solved. At impact-oriented pitch events—those geared toward major social and environmental challenges—these problems are portrayed as particularly urgent and enormous; otherwise the problems generally amount to a significant friction in business or everyday life that if erased would make work or leisure easier, more comfortable, or fun. The pitch then moves to the specific technological solution on hand, describing its unique character and groundbreaking technology. This portion is followed by projections about the size and scope of the market for the product, projections that are often bloated to the point of incredulity. The pitch always ends with brass tacks, when pitchers impress upon their future funders that they have the team and expertise to develop the technology, including alliances with incumbent corporations and a plan to assure profitability within a few short years. In this part of the pitch, many entrepreneurs emphasize how many patents they have pending or have already obtained. This is a signal to potential funders that their idea will not be immediately replicable and therefore a potential money-maker if the product is taken to market. Following the pitch, judges, some of whom are funders, pepper the entrepreneurs with either coaching or questions in efforts to vet for the most promising. This style has clearly bled into popular culture, with shows like *Shark Tank* virtually replicating this approach to product marketing as a form of entertainment. As you will see, it has bled into university life as well.

The structure and time constraints of the pitch demand that the entrepreneur strip down their ideas to the easily digestible. Pitchers have no time, much less inclination, to describe the problem and say how they know it to be a problem; to explain why their solution is the most optimal or whether the problem even warrants a technological solution; to justify the size of the market when many others are pursuing similar ideas; or, for that matter, to make sense of how they want to cozy up to the very businesses they claim to disrupt. In other words, the pitch requires the omission of doubt and complexity. Given how much the ecosystem prepares entrepreneurs for the pitch, as if the pitch is as good as the creation, it thus appears that the ecosystem itself discourages all such reflexivity. Instead it instills a solutionist mentality in which, to paraphrase Morozov, the answers are reached before the questions are asked.[43]

In the interviews we conducted with Silicon Valley entrepreneurs, as part of our research project, we were repeatedly struck by entrepreneurs who were unable to explain how they researched problems or arrived at solutions. More often than not, they repeated the material they presented at the pitch or posted on their websites. As the handbooks and mentors instruct them: "You're always pitching."

## The Problem with Silicon Valley's Approach to Solutions

It would be one thing if Silicon Valley was continuing to make gadgets or speeding and smoothing processes. Yet Silicon Valley's indisputable track record in fostering world-changing innovations has made it *the* model for innovation and change, even for addressing highly challenging and complex social and ecological issues. The idea that the tech sector can solve the so-called wicked problems of our day—referring to those that are highly complex, interrelated, and often intractable—has been embraced by public and private figures alike. For their part, both successful and aspiring techies believe they have what it takes to make the world a better place. So enamored with their own approach, they often bypass existing efforts to create their own.[44] The problem is that the Silicon Valley model is not only a product of its political economic and cultural context; it is also constrained by it, as promised profitability, aversion to social change, and intellectual simplicity limit the range of solutions even under consideration.

Sociologist Jesse Goldstein's account of cleantech, the domain that was supposed to usher in Silicon Valley's new dedication to environmental impact, illustrates how the funding mechanism in particular constrains possibility.[45] Describing what he calls nondisruptive disruption, Goldstein posits that the limits of the Silicon Valley model rest on the dual imperatives of impact and profit. Through his research he found that cleantech entrepreneurs were both inventive and serious about their aspirations for planetary improvement. For their part, venture capitalists largely supported entrepreneurial efforts at impact, indeed judged entrepreneurs on their perceived ability to deliver it. At the same time, venture capitalists imposed discipline on entrepreneurs that effectively tempered the latter's ambitions. Funders demanded that new clean technologies reach a return in a few short years and dismissed entrepreneurs who put too much emphasis on their creativity relative to profits or market share. By venture capital standards, then, the most successful entrepreneurs were those who could walk that fine line, promising an impact greater than making money but showing a willingness to submit to the dictates of capital. You can see that line being walked in the pitch, as entrepreneurs seesaw between the big problems they will solve and industries they will disrupt, and the business plans they will follow and incumbents to which they will submit.

Under these circumstances there is one surefire way to demonstrate impact, according to Goldstein: to highlight the wrongs of the status quo. Articulating the damage of a highly polluting or abusive industry necessarily makes delivering any alternative an impact. So does depicting an imminent crisis, like the need to feed the coming ten billion. There is a reason that start-ups amplify problems in those pitches! But the solutions often end up being the low-hanging fruit, the stuff that's easy to fix with little sacrifice and thus with little rearrangement of current ways of living and social structures. So you get hybrid cars rather than vibrant public transportation systems, carbon-capturing gadgets rather than reduced air travel, recycling rather than the substantive curtailment of waste. These are not only classic techno-fixes; many are either highly mundane or incremental, hardly the earth-shattering technologies that will change the world for the better. Goldstein, in short, makes a convincing case about how the Silicon Valley ecosystem described in this chapter yielded quite modest solutions in the cleantech sector, despite its genuine aspirations to make the world a better place.

In chapters 3, 4, and 5, I explore whether Silicon Valley's forays into agrifood have served up something better. First, however, it is important to provide more context about food and agriculture itself as a domain ripe for transformation. Chapter 2 looks at earlier efforts to address the challenges of food and agriculture with technology. One project in particular, global in reach and closely aligned with postcolonial development, set an early precedent for today's agrifood solution-makers, predating neoliberalism. Bearing lessons all but ignored by Silicon Valley changemakers, the Green Revolution provides an exemplary case of how a misconstrued problem, an urge to avoid politics, and a will to improve led to a solution with consequences that remain controversial to this day.

## 2  Agrifood Solutions before Silicon Valley

In 1970, Norman Borlaug, a geneticist at the Rockefeller Foundation, won the Nobel Peace Prize—one of the first of many prizes and awards that Borlaug was to receive for his contributions to the Green Revolution. The Green Revolution, which took place roughly from the 1940s through the 1970s, was a concerted effort to increase food output in the developing world. Far from an environmental reckoning—it was not an effort to "go green"—it primarily involved the application of traditional plant-breeding methods to make the staple crops of wheat, maize, and rice—those that fed much of the world's population—more productive. For many, especially at the time, the increase in agricultural productivity presented an unprecedented solution to hunger in the developing world. Witnessing its consequences years later, however, the jury was split. Writing in 1997, the *Atlantic* columnist Gregg Easterbrook deemed Borlaug responsible for ensuring that global food production had expanded faster than the human population throughout the postwar era, "averting the mass starvations that were widely predicted." Believing that Borlaug never received the notoriety he deserved, Easterbrook further lauded "the form of agriculture that Borlaug preaches" for "prevent[ing] a billion deaths."[1]

The Green Revolution's detractors came to different conclusions, doubting its contributions to ending hunger and recounting its many undesirable consequences. It is not as though the problem of hunger demanded an approach centered on increasing plant yield. Other options were available, not least of which were those focused on redistribution of land, food, or wealth. But promoters of the Green Revolution held to a particular framing of the problem that rendered it technical. Simply put, they pitched it as not enough food relative to growing human populations, for which increasing food production through technology was the answer. Focused on improving the biology of the plant through technoscience, the Green Revolution was thus a paradigmatic example of a techno-fix for agriculture: a purposefully technical approach intended to avert fundamental social and economic changes. At the same time, it was guided by the will to improve. The Green Revolution's champions sincerely believed they were fixing a problem, even as they subverted the more radical social approaches some publics were calling for. And it contained aspects of solutionism as well, as it was driven by people whose particular expertise shaped the approach.

That the Green Revolution continues to serve as an exemplar of how to fix food, albeit often unwittingly, is not cause for celebration. Often acting with seemingly little if any awareness of the conceptual underpinnings, historical origins, and consequences of the Green Revolution, today's food innovators continue to offer solutions with potential to replicate the mistakes of the past. As the mother of all agricultural techno-fixes, the Green Revolution should instead serve as a cautionary tale.

## The Ghost of Malthus

We can locate the conceptual underpinnings of the Green Revolution in Borlaug's Nobel lecture. In that speech Borlaug balanced his enthusiasm with the "vital role that agricultural improvement would play in a hungry world" against the ongoing specter of population growth. As he put it:

> [T]he Green Revolution has won a temporary success in man's war against hunger and deprivation; it has given man a breathing space. If fully implemented, the revolution can provide sufficient food for sustenance during the next three decades. But the frightening power of human reproduction must also be curbed; otherwise the success of the green revolution will be ephemeral only.[2]

In making this statement, Borlaug was rehearsing a thesis made famous by the economist Thomas Malthus, a cleric with the Church of England who preceded Borlaug by about 150 years. Malthus first established the thesis of population outstripping resources through his *Essay on the Principle of Population*, published in 1798.[3] He based it on two postulates: "food is necessary to the existence of man," and "the passion between the sexes is necessary, and will remain nearly in its present state." We might word these differently today, by acknowledging, among other things, the diversity of sexes, genders, and sexual interests. Still, the idea that people like to eat and have sex is not groundbreaking. The controversy in Malthus's logic lay with the assertion that "the power of population is indefinitely greater than the power in the earth to produce subsistence for man." Here he noted that population can always grow, but earth contains a finite amount of land.

I imagine some readers may be wondering what exactly was wrong with the argument. After all, it is hard to argue with the proposition that population can always grow but that the earth's land is finite. In fact, many political economists of Malthus's day, including Marx, theorized the implications of the world's limited land base. Yet others saw the possibility of agricultural improvement and agrarian reform as a way to mitigate hunger or starvation and argued for various means to improve standards of living. For Malthus, though, any improvements would just incentivize more population growth. "I see no way by which man can escape from the weight of this law which pervades all animated nature," he wrote. "No fancied equality, no agrarian regulations in their utmost extent, could remove the pressure of it even for a single century." Indeed, Malthus's pessimistic ideas about human improvement led him to conclude that nothing could be done except letting the poor, whom he saw as lacking moral restraint, die.

The ghost of Malthus looms in many conversations and policies today, including the idea that population growth is the basis of environmental problems. Skeptics of this position, who find other explanations of environmental problems, from "overconsumption" to capitalism, have accused some branches of the modern environmental movement of being neo-Malthusian. For our purposes the takeaway is that proponents of the Green Revolution accepted the Malthusian idea that insufficient food in relation to population growth was the cause of hunger. They similarly rejected solutions to hunger involving any sort of redistribution of land, wealth, or food. However, Green Revolution promoters were more optimistic than Malthus about the possibility of mitigating hunger, albeit in a very specific way: the seemingly apolitical techno-fix of increasing productivity. Their bounding of the problem to reject redistributive solutions must be understood within the historical context of the Green Revolution.

## The Green Revolution and Its Critics

The origins of the Green Revolution date back to the 1930s, when the Rockefeller Foundation took an interest in improving health in Mexico as part of the United States' efforts to keep Latin America in its political orbit. The US secretary of agriculture at the time, Henry A. Wallace, convinced the head of the foundation to put its emphasis on agriculture, "so as to make sure that food expanded as fast as people."[4] That said, when the foundation expanded its agricultural research in the 1940s, its focus was less on "overpopulation" than in modernizing agriculture in the region. During the 1940s a new government had reversed Mexico's commitment to land reform, a key promise of the Mexican Revolution of 1910. The Rockefeller Foundation and the United States Department of Agriculture (USDA), keen on restoring export markets and reinstating past patterns of land ownership that favored Anglo-Americans, were particularly influential in convincing the Mexican government that peasant sectors were inefficient and that agricultural modernization could help subsidize industrial growth. Their influence helped usher in the reinvigoration of US agricultural research in Mexico, but sounded the death knell on the possibility of land reform. In 1943 the Rockefeller Foundation, in cooperation with the Mexican government, set up a research project to improve local wheat and maize varieties and brought along Norman Borlaug to lead the effort. With Borlaug's expertise in plant genetics, a solution that lay with enhanced productivity rather than agrarian reform became a near inevitability.[5]

Mexico became the showcase for extending the Green Revolution elsewhere, including the rest of Latin America and Asia.[6] China's 1949 socialist revolution made the country impermeable to Western capitalist influence, and the 1945 nationalist revolution started Vietnam down a similar path. During the Cold War the Non Aligned Movement—leaders of a large group of former colonies, including the highly populous countries of Egypt, India, and Indonesia—wished to separate from both the USSR and the United States. (This, by the way, was the actual origins of the term "third world," which in its earliest meaning did not refer to degrees of poverty.) As this largely nationalist movement threatened to separate countries from US politics and markets, trustees representing US interests wanted to keep these former colonies within the US sphere of influence. Moreover, they doubted that peasant agriculture could feed these highly populous countries and worried that strife from food shortages and famine would foment further "red" revolutions. In focusing on insufficient food, however, these trustees neglected to consider the reasons that colonialism had left many countries impoverished, such as the confiscation of the best land by elites, or in the case of the British Raj in India, a tax system that had drained the countryside of revenue. But rather than address the social origins of the problem head-on, which might have led to a response involving the redistribution of land or wealth, they turned to the technical solution of productivity-enhancing technology. That Green Revolution technologies would maintain or even enhance the status quo reveals just how political an allegedly apolitical techno-fix can be!

The Philippines became the first center of Green Revolution activity in Asia. In 1960, with the support of the Ford and Rockefeller Foundations, the government of the Philippines established the

International Rice Research Institute (IRRI). A year later the minister of agriculture in India invited Borlaug to begin a program there that would focus largely on the water-rich Punjab. Eventually the IRRI established offices in seventeen countries. Technically, the Green Revolution largely focused on the development of highly productive crop varieties. IRRI's primary contribution was the development of "dwarf" rice varieties, which would minimize the energy plants put into growth. Scientists bred other traits as well: standardized heights to minimize the problems of shading, photoperiod insensitivity to reduce the effects of seasons (allowing two crops per year), rapid maturation to speed growth, and resistance to pests and disease. The technologies disseminated widely, and by the late 1990s more than 80 percent of Asian rice areas were growing varieties born of the Green Revolution.[7]

Many applauded the results of the Green Revolution. For the fifty years after 1960, the production of cereal crops tripled, with only a 30 percent increase in land area cultivated, significantly outpacing population growth.[8] According to a 2012 review of the academic literature, the Green Revolution "contributed to widespread poverty reduction, averted hunger for millions of people, and avoided the conversion of thousands of hectares of land into agricultural cultivation."[9] Some attribute the lack of famine in India in the 1960s to IRRI's semidwarf varieties.[10] Today, many development advocates want to replicate its success in Africa. Yet over the years the Green Revolution has also drawn criticism from all over the world, and not only from social scientists and activists. Even the Ford and Rockefeller Foundations, which initially funded it, as well as the World Bank have distanced themselves from Borlaug's legacy.[11]

Some shed doubt on whether the high-yielding varietals actually contributed to more productivity, especially since more land had been brought into production. Virtually all detractors agree that the Green Revolution was rife with negative environmental and social consequences. To produce those high yields, farmers had to obtain and use a package of inputs, including seeds, nitrogen fertilizer, pesticides, and water. Use of these inputs was costly and contributed to a number of ecological problems: groundwater depletion, pest buildup, micronutrient depletion, and disease susceptibility.[12] Investments in this package encouraged monocropping too, which not only attracted disease and pests but contributed to the loss of rice varietals that could be useful in the future.

Critics have additionally questioned whether increasing yield actually contributed to a decline in poverty and hunger. It certainly didn't help the many farmers who became landless or indebted and therefore no longer had the income to purchase adequate food (see the Sen analysis below). Confounding the issue, several of the governments that had pursued the Green Revolution developed complementary programs to address hunger. For instance, India implemented a public distribution system that involved purchasing grain from farmers and providing it cheaply to people in cities. Mexico created a similar system of subsidizing corn and tortillas by purchasing corn and beans from producers. Mexico's system channeled the surplus either directly to urban consumers or to the corn flour and tortilla industry. These sorts of food assistance programs were crucial for actually getting food to the poor; increasing output isn't much of a fix if crops are not being converted to food and distributed to those in need. In other words, it took social organization to effectively address hunger.

Given the Green Revolution's multifarious negative consequences, it is tempting to read it as a cynical plot. When I ask my students what they know about the Green Revolution, they commonly answer that it was a program to increase profits for the corporations which in turn ruined ecologies and left people landless. They have a hard time imagining that it was motivated by good intentions, much less that Borlaug was deserving of a Nobel Prize. But I think they confuse intentions with effects. That the United States had geopolitical interests in the Green Revolution is indubitable.[13] Rockefeller and Ford were trying to save the developing world from socialism on behalf of American capitalism. The Green Revolution surely helped the bottom lines of corporations as well.

But their motives were *also* altruistic. Many of the institutions that led the Green Revolution were nonprofit organizations (NGOs) and foundations that genuinely desired to address hunger and poverty in the so-called third world. Indeed, I think reading the Green Revolution as only a cynical plot on behalf of corporations misses the significance of the will to improve. Those behind techno-fixes may well be making money, but they also act as trustees, believing they have the capacity to fix things that others lack. The fault in their approach was in rendering the problem technical and amenable to what they could offer (higher yields) so that other avenues of addressing the problem were muted. The problem of hunger was real, but increasing productivity as the sole aim was not the answer.

## The Sen Response

If the Green Revolution was a quintessential example of the problem with solutions, the question, then, is what a response might have looked like. You can see the contours of response in the work of the economist Amartya Sen, whose analytical framework provided a robust counterpoint to neo-Malthusian thinking and its focus on increased output as a solution to hunger. Born in 1933 and still living as of this writing, Sen is a Nobel Prize winner in economics and has earned ninety honorary degrees from around the world. As a nine-year-old boy, Sen personally witnessed the Bengal famine of 1943, during which three million people perished—a famine that also played a large role in spurring early Green Revolution efforts. A fraught period during the 1970s also inspired his writing on hunger and famine. The world had seen famines in Ethiopia and Nigeria, rising food prices, and a general fear of scarcity. Through careful research and analysis about the character of the problem, Sen made the case that hunger did not stem from insufficient food output.[14]

Forwarding what he called a "capabilities" approach, Sen formulated the idea that chronic hunger arose from the lack of an ability to reliably gain access to, and command over, food. Specifically he brought focus to the social relationships of individuals, households, and communities that affect how people obtain food and how those relationships can change. For Sen, factors such as citizenship status (i.e., relation to the state), occupational status and wage levels in relationship to prices, and even the traditional rights within the community all could mediate access to food. Notably, he did not specify one particular cause of hunger or famine. Instead, Sen provided a set of tools for analyzing any given hunger situation.

Two concepts were at the core of the toolbox. Sen referred to "endowments" to describe an individual's original capabilities and resources. For example, peasant farmers' endowments may be their land, labor power, and a few other resources such as farm implements, while workers may have only their own bodies (labor power) as endowments. Sen used the term "entitlements" to refer to what an individual with a particular endowment can legally acquire. They may be able to grow food (typical of subsistence farmers), sell and buy food (typical of market-oriented farmers), sell their labor and buy food (typical of workers), or obtain food from their families, communities, charities, or the state. None of these are sure things, though. Wages may be too low relative to prices. The prices farmers receive for cash crops may not be sufficient to buy food. Or government food support may be insufficient or nonexistent if you are, say, an undocumented worker.

From Sen's framework, you can get a sense of how entrenched political and economic structures can affect hunger, because they often determine the distribution of entitlements: citizenship laws, prevailing wages, access to good land for growing food, and access to markets to sell food. Ideology can matter, too. In some societies, for example, tradition holds that girls in the household eat last, contributing to their hunger. Even geographic access to purchase or obtain food can be seen as a kind of entitlement that some may lack. Sen developed this framework out of his analysis of famine. Through his research he found that many famines occurred in periods of growing food output. He further noted that many famine victims, defined by those who sought relief or died, were often those who produce food. Their having access to land did not guarantee freedom from hunger. In fact, in Sen's analysis of four famines in the twentieth century—the Bengal famine of 1943, the Ethiopian famines of 1973 and 1974, and the Bangladesh famine of 1974—he found that occupational groups closest to food production accounted for the largest number of famine victims. These included rural laborers, farmers, pastoralists, and fishers. Why? Others had little cash to purchase food. Producers unable to sell their produce, or at very low prices, would find their entitlements insufficient to buy food. In short, Sen showed that famine or hunger were best understood as a failure or insufficiency of entitlements, not an insufficiency in food output.

In recent years scholars with affinities to Sen's work have shifted the focus from "failure to access" to "failure to respond" in the case of famine.[15] These scholars emphasize that most famines are preventable political failures. In this "new famine thinking," the role of the state stands at the center of the discussion. The prominent example is the 1948–52 famine during China's "Great Leap Forward" in which the Chinese government undertook a program of rapid collectivization of farms and industrialization. To feed the cities, the Chinese state purchased something like 30 percent of rural grain output, leaving very little in the countryside. As a result, tens of millions of people died. Other examples include the Irish Famine of the 1840s, caused by the unwillingness of the British government to intervene when potato blight struck, or the Biafran famine of the late 1960s, precipitated by the Nigerian government's withholding of food aid to Biafra. The Nigerian government did this to punish the Biafran people for attempting to secede from Nigeria once they found oil in their region. Civil wars or untoward invasions are often a cause of famine and hunger too, because they often disturb food production. Witness the rise in food prices circa 2022 because Ukraine's robust wheat production was curbed by the Russian invasion.

Sen's framework, inclusive of the new famine thinking, does not provide a clear causal theory of hunger and famine. For example, both economic booms and busts can bring about famine. Busts deprive people of purchasing power from lack of income while booms price goods too high. In other words, entitlement failure can happen in any number of ways, just as food security can manifest from multiple means. Yet it does provide conceptual tools for formulating more robust response. We can take from Sen an invitation to investigate the specificities of any given hunger situation rather than make the standard assumption that it lies with insufficient output. The outcome of such an investigation, the problem analysis, presumably would point the way to a response appropriate to the problem.

Despite certain shortcomings in Sen's framework, including its tendency to focus on individual capabilities rather than structural forces, he has had a tremendous influence on thinking about hunger, including shifting the frame from hunger to food security.[16] The creation of the aforementioned public distribution in India, which bought food from farmers to support urban dwellers, was based on Sen's analysis of famine. The current focus of food desert activism on "access" owes much to Sen. We can look to Sen when we say, often without attribution, that the problem is food distribution not production, and that a core way to address this distributional problem is to provide incomes or more robust food assistance to poor people. Sadly, these lessons have eluded many agrifood solution-makers, from university agricultural researchers to techie entrepreneurs to students wanting to effect change, who continue to pursue agricultural productivity as a solution to hunger.

## Why Productivity Will Not Help Farmers Much Either

Despite well-trodden critiques, the Green Revolution has served more as an example than a cautionary tale for today's innovators in food and agriculture. Indeed, it is striking how ubiquitously they invoke the ghost of Malthus in claiming their inventions will "feed the coming ten billion." Go to an agrifood pitch event and pitch after pitch will begin with this thought. When pitchers are not reciting the line about feeding the ten billion, they sometimes use another go-to line: they want to make farmers more profitable. But guess what? Productivity-enhancing technologies have not been great solutions for farmers either, and this too has a history.

Consider the United States and its experience with "productivism." Productivism has been described as "a commitment to an intensive, industrially-based and expansionist agriculture with state support based primarily on output and increased productivity."[17] In the United States a focus on productivism began with the New Deal, carving an approach that was replicated around the world, including in the Green Revolution. To set the scene, the last third of the nineteenth century and the first third of the twentieth century had seen series of booms and busts in agriculture. These were largely due to colonial expansion as new parts of the world were brought into export-oriented crop production. With recurring gluts, agricultural prices would fall, threatening the livelihoods of farmers; with ensuing shortages, prices would rise and encourage ever more expansion. This cycle eventually led to the international Great Depression. At the time many Americans were hungry despite the vast amounts of grain that US farmers had produced, largely left unsold.[18]

And so the United States pursued a range of policies to bring stability to crop markets and support farmers, and, as a by-product, to curb urban hunger. Most famously, it created the Commodity Credit Corporation (1933), a program that encouraged farmers to store grain with the government whenever prices fell below the costs of production. Farmers would receive a loan from the government in return for the grain to discourage them from dumping grain into a weak market. Theoretically, farmers would pay back the government when prices were good and sell the grain at a profit. If prices didn't recover, the government would retain the grain and farmers could keep the borrowed money without recourse. In effect, this became a minimum price support for farmers and resulted in the federal government owning huge volumes of surplus grain. This government surplus led to one of the first efforts to kill two birds with one stone: supporting farmers while providing the surplus food directly to hungry workers. Indeed, this preceded the programs taken up in India and Mexico.

This was not the only policy, however, that the United States pursued to manage excess production. It also paid farmers to shift land into soil conservation and implemented marketing quotas and acreage restrictions. All of these were effectively supply management strategies, efforts to support farmers' prices and keep them in business by keeping crops off the market. At first these programs applied to only a few "basic" commodities: wheat, corn, hogs, cotton, rice, and tobacco; later the programs expanded to other crops but never reached the specialty crops of fruits, nuts, and vegetables.[19] Alas, the effect of these programs veered from their intentions. As policy discouraged farmers from expanding acreage, they learned to intensify instead. "Intensification" in agriculture refers to various means to increase productivity on any given piece of land. In other words, rather than increasing production by expanding your acreage, you do so by maximizing productivity on a per-acre basis. How did farmers undergo intensification? Generally by adopting the technological solutions that were being made available to support that very thing.

Providing a model for the Green Revolution, these technologies included breeding plants and animals to increase the size or number of units, enhancing overall output, or accelerating the production cycles. Breeding and eventually genetic engineering became the primary approaches to make plants and animals grow faster, bigger, or reproduce more abundantly. The quintessential example was the introduction of hybrid corn. Hybridization involved crossing two inbred strains. It significantly increased yield and since sterile females need no detasseling, farmers could harvest their corn by machine. The downside is that seeds from hybrid corn would not work well for the next generation and so to utilize these traits, farmers had to purchase seeds every year. An example in livestock breeding is that animal scientists bred broiler chickens to have more breast meat, to the extent that modern chickens can no longer walk, and cut their maturation times from seventy to fifty days.[20] Nutrition and pharmaceutical treatment, including the prolific use of antibiotics and growth hormones, also hastened and amplified the growth and reproductive cycles of livestock, while synthetic fertilizers did that for crops.

By minimizing the risks of plant and animal loss to disease and pests, a range of mechanical and chemical technologies also promoted intensification. Following World War II, this included the use of DDT, which had formerly been used during the war as an effective malaria control. After the war DDT was applied to plants until it was eventually banned in the 1970s for its carcinogenic qualities. And, of course, the increased use of machinery, whether for preparing fields, harvesting grain, or milking cows, saved farmers labor costs and also contributed to overall productivity. In short, New Deal farm programs, originally intended to curb supply for farmers, actually incentivized massive increases in farm productivity.

None of this was particularly good for farmers and ranchers. Yes, they appreciated the labor-saving technologies that minimized their dealings with the cost and occasional recalcitrance of farm laborers. Otherwise, intensification undermined agricultural prices, owing to a phenomenon called the "technology treadmill." As explained by the agricultural economist Willard Cochrane, who first coined the technology treadmill concept, "early-bird" farmers, those who adopt a new and improved technology, see a reduction in per unit costs. At first, the increased output of a few farmers has a negligible effect on prices, but as more farmers adopt the technology, the supply on the market increases, causing prices to fall. With widespread adoption, prices eventually fall to the point that all gains are eliminated. And "laggard" farmers, those who do not adopt, experience losses, as their expenses end up surpassing existing prices.[21] The dynamics of the technology treadmill pretty much predict how the US farm sector became increasingly consolidated over the twentieth century, especially in basic commodity crops that were more amenable to large-scale mechanized farming. With new technologies, laggard farmers or others who just experienced bad luck went bankrupt, often saddled in debt. More aggressive farmers, who were able to snatch up land and other productive assets, grew increasingly larger. But even the more successful farmers who remained had to operate at increasingly larger scales, maximize efficiency, and accept low profit margins just to stay in business.[22]

Something very similar happened during the Green Revolution: accessing key inputs contributed to massive differentiation of farmers. Green Revolution promoters claimed that these technologies would be scale-neutral, such that farmers of any size could use them. They weren't. Farmers needed capital to buy the seeds, fertilizers and pesticides, and land with private wells to have adequate irrigation. Those who had the capital and well-irrigated land made those investments and had good results. Even these successful farmers had to purchase these inputs from foreign corporations though, making them vulnerable to exchange rates, inflation, and international debt issues. But many farmers couldn't benefit by Green Revolution technologies because they farmed in drought- or flood-prone conditions or simply did not have the capital. Most farmers lost their land and either became landless laborers for other farmers or fled to the cities.[23]

In other words, more technology did not make farmers more profitable. To the contrary: the solution of more productivity became the problem.

## The Persistence of Techno-Optimism

Despite the well-documented downsides of the Green Revolution and productivism in general, the idea persists that even more agricultural intensification is the answer. What's more, those promoting that view call themselves technological or environmental optimists. In his essay "Can Planet Earth Feed 10 Billion People?," Charles Mann refers to "technological optimism" as "the view that science and technology properly applied will let us produce our way out of our predicament."[24] Gregg Easterbrook, the writer introduced at the beginning of the chapter who championed Borlaug's work, is a self-described environmental optimist, adhering to that precise idea.

I find this self-ascribed technological optimism remarkable. It rests on the persisting assumption that population is the essential problem when the far more trenchant problem is rampant inequality in access to resources. It is particularly questionable since the world's rate of population growth is slowing considerably and may well peak at ten billion. As such, I would argue that such optimism is in fact very pessimistic. Techno-optimists appear to imagine no other responses than the solution of increasing output—or, for that matter, no other uses for agricultural technology other than to increase output. It is as if the techno-fix is the only game in town, relegating other responses from redistribution of land to guaranteed incomes to more robust food assistance programs as pie in the sky—too politically challenging to be considered and not as immediately tangible as a technological solution. Technological solutions have that appeal, but we should not forget that before and during the Green Revolution such responses to hunger *were* on the table. Land reform especially was a widely supported and reachable political program of many former colonies at the time, to the extent that supporters of the Green Revolution aimed to thwart it.

So we ought to be cognizant of current-day efforts that might be getting short shrift because the techno-fix of more productivity rocks fewer boats. Especially if we heed Sen and others in his wake, the problem of hunger really does require a political response that can address its social causes. Imagining a more appropriate response to be impossible is not very optimistic at all. And still, this very technological optimism is rampant in the Silicon Valley tech sector, which is now bringing to bear its expertise and business model to the problems of food and agriculture. While initially stumbling into food, Silicon Valley has since crowned itself heir to the Green Revolution, taking on not only the challenge of food security but a range of health, animal welfare, and environmental problems implicated in food production. While techies hope to solve problems beyond hunger, they rehearse the need to feed the world, as if nothing was learned from these past moments.

How did Silicon Valley, so different in its institutional framework than that of the Green Revolution, become the standard-bearer for agrifood innovation in the first place? I turn to that in chapter 3.

## 3  Silicon Valley Bites Off Agriculture and Food

In 2019, as part of the research project outlined in the introduction, I attended a pitch event typical of Silicon Valley investment and start-up culture. As these events go, it was a low-key, low-entrance-fee affair, designed to generate enthusiasm for techies flirting with the agrifood tech space. As such, the event provided an opportune moment to learn why food has become of interest to the tech sector. It didn't disappoint.

The evening began with an hour of networking, with folks wearing handwritten name tags and mingling over wine, sodas, cubed cheese, and water crackers—not the lavish food displays I'd witnessed at similar events. As soon as my colleague and I parked ourselves at a high-top table, attendees made their way to speak with us. Looking quizzically at our name tags with university affiliations, they would enthusiastically introduce themselves as if we had something to offer them. When we asked attendees what they were doing at the event, the reply was a consistent "I love food." While some participants followed with "I'm just checking it out," many more elaborated on their interests. One person, for example, talked to us about how he left a tech job, working as a programmer, because it was too dehumanizing. He wanted to start a company that would "give people something they could love." He decided on a "high-quality artisanal dessert" that he said, following an incorrect explanation of the term "terroir," reflected the "terroir" of San Francisco and told us he was using all natural ingredients. We asked what any of this had to do with tech. His answer was that he was using some high-speed cooking technology. Cooking is like coding, he averred, because you have to create each piece and then put them together, just like combining ingredients for a recipe.

Another attendee identified himself as a "food technologist" and said he was from a food tech start-up club at a nearby city college. When we asked why food, he waxed rhapsodic about his grandmother's tzatziki and the importance of authenticity. "Tzatziki is made in a specific way; if you change an ingredient, it's not tzatziki," he exclaimed, before asserting that the San Francisco food scene lacks authentic international foods. But then, in startling contrast, he went on to rave about automation, sharing a video of a street kiosk in San Francisco that houses a coffee-making robot. "No lines," he said, "the food industry needs to eliminate as much as possible the human factors."

Two more attendees came over, these two investors, one with the title "angel" on his name tag, referring to a kind of investor who graces early start-ups with funding, usually in exchange for a share of equity. The angel told us that he had been to seventy countries where he would eat and learn to cook the local fare, punctuating the point with a proclamation of love for the food of his ancestors and a desire "to imitate that." Without missing a beat, he shared that during his recent attempt to become vegan, he discovered a love for the Impossible Burger. Professing his adoration for natural foods and his skepticism of genetically engineered foods, the angel expressed surprise to learn (from us) that the Impossible Burger contains an ingredient produced through genetic engineering. Soon after, the conversation shifted, and the second investor asserted that agriculture is not making good revenue, asserting his interest in getting producers to get "respectable yields."

That evening it became very clear that one of the big appeals of the agrifood space for techies and investors is that it is more meaningful, fun, and delicious than coding. It also became evident they believed that tech could bring solutions—and profit—to food and agriculture that heretofore were apparently missing. Never mind that the innovations that interested them (robots, Impossible Burgers) were contrary to the very things they love about food (artisanal crafting, authentic foods)—or that some of these innovations were commonplace (plant-based burgers, technologies to increase yields). The question is how food and agriculture became a space for Silicon Valley innovation and investment to begin with.

## Origin Story

In 2013 the large and much scorned biotechnology company Monsanto, which has since been acquired by the even more massive Bayer Corporation, purchased a little known start-up called Climate Corporation for almost a billion dollars. Climate Corporation had been created by two former Google employees and backed by a number of high-profile venture capital firms. The latter included Founders Fund, headed by Peter Thiel who, among other things, cofounded PayPal and provided early funding for LinkedIn. Monsanto's purchase was a provocative move for an erstwhile chemical company that built its reputation (good and bad, depending on your perspective) out of bioengineering seeds to be compatible with the use of its highly disparaged weed-killing formula Roundup, containing the controversial chemical glyphosate. Monsanto was also famous for suing farmers for saving their own seed or having Monsanto's bioengineered seeds inadvertently drift into their fields. Climate Corporation offered a machine-learning technology to monitor and predict climate change. Licensed to farmers as a platform, it would allow farmers to view data on climate conditions and aid in managing climate risk—something like the promise of OpenAg's food computers. Some in the business press suggested that Monsanto's recent interest in climate would burnish its reputation.[1]

This deal warrants attention for several reasons. One is that it involved the acquisition of a supposedly disruptive company by one of the bad boys of agribusiness; another is that it involved the uptake of a digital technology, developed with typical Silicon Valley expertise, from a company whose mainstays have been biological and chemical technologies; a third reason of utmost significance is that by the accounts of many of those we interviewed, Monsanto's purchase marked the advent of agrifood tech as a discrete arena of business activity and investment worthiness—what I'll refer to as a sector. Up until this acquisition, start-up activity in food and agriculture was dispersed and unremarkable. But all of the sudden, agriculture and food appeared as the next big thing for Silicon Valley to make its imprint and fulfill its dreams of worldly impact. Crucially, it came right on the heels of what turned out to be a bust in cleantech. After once enjoying venture investors showering money to the tune of $25 billion between 2006 and 2011, what is now called CleanTech 1.0 went by the wayside as funding dried up. Apparently these earlier investors hadn't accounted for the deep technical risk, long development timelines, and capital intensity associated with cleantech investing.[2] The 2008 financial crisis didn't help.

The Monsanto deal signaled that the big boys were interested in tech, but what else made agriculture and food appear ripe for Silicon Valley to bring its unique style of innovation and solution-making? One set of factors revolved on the nature of food. Food is a domain in which human life and livelihood are the most deeply entwined with multispecies' planetary futures. The planet's grandest challenges of climate change, environmental sustainability, human health, and poverty all implicate current modes of food production and distribution. Problems more specific to this sphere—inhumane and unhealthy livestock production, toxic chemical-intensive crop production, soil depletion, livelihood struggles for farmers and farmworkers, food insecurity—are also major areas of concern.

Food is also obviously one of life's basic necessities, something that everybody needs, giving the sense that there are no wrong steps in trying to meet those needs. And food is deeply material. At a time that virtual reality and mediated experience (which are still dependent on the material world of circuits, plastics, metals, and labor) increasingly preoccupy our time and imaginations, food and its production have a concreteness that is irreplaceable. Even if you play video games all day long (on your material laptop computer), when you eat food you are bodily connected to dirt, water, plants, and animals in ways different than the connections you feel to whatever is on your screen. The concreteness of food lends extra moral and emotional weight to efforts to transform it. If Silicon Valley seriously wanted to have impact, what better place to look?

A second set of factors stemmed from a sense that the techies could do better than previous efforts to transform food. As many food activists have found, it is easier to critique industrial corporate food than to change it. Altering farm policy, enhancing regulation of toxic or unhealthful farm inputs and food additives, breaking the power of highly consolidated agrifood corporations, organizing for living wages and incomes for food producers, and ensuring that people have enough nutritious food to eat take massive organizing efforts and a dramatic change in political will, not to mention a functioning democracy. Solving food problems politically can appear pretty damn daunting.

One way activists have gotten around some of these political challenges is through the development of alternatives. These alternatives are wide-ranging and include organic and regenerative agriculture, rotational grazing, and agroecology; whole and natural foods, artisan foods, cooking education, vegetarian and vegan diets; and farmers' markets and community supported agriculture, community gardens, food hubs and cooperatives, and community-owned food marketplaces. Despite widespread interest in and public and scholarly discussion about these alternative approaches, the take-up of these practices has been limited. Organic agriculture, for example, by far the most developed and widely recognized of all these alternatives, still comprised less than 1 percent of farm acreage in North America in 2022.[3] Although part of the reason these already-existing practices toward agricultural sustainability and food justice have remained in the margins is they have not received the same funding and hype as the solutions emanating from the tech sector, these efforts have nevertheless left the appearance of a gap—a gap that the tech sector could fill by bringing its own brand of change-making to the world of food.

A third set of factors drew from Silicon Valley's cultural and political economic logics. Agriculture and food appeared dominated by some of the most sclerotic, entrenched incumbents: the "big food" and agribusiness companies that everyone loves to hate, from Tyson to Monsanto. If disruption was in the offing, food and agriculture looked to be unusually ripe for it. For that matter the domain seemed seriously "underinvested," as recounted many times in interviews. Given that food and agriculture have long been penetrated by large corporations, I came to understand that "underinvestment" meant that venture capital had not yet made its mark. And so beginning around 2013, a number of facilitating organizations came on the scene to organize conferences, hold pitch events, form networks, and create platforms to make agrifood tech a thing. Since then, agrifood tech has attracted substantial investment. Global data collected by AgFunder, a venture capital firm that follows investment deals in the sector worldwide, reveals a whopping $51.7 billion in total funding for the year 2021, representing 3,155 deals. Significantly, this figure represents an exponential acceleration of investment, increasing more than twentyfold since 2013, which recorded $2.3 billion in deals.[4] Some agrifood tech companies have received much more than venture capital interest. The Boston-based company Indigo Agriculture's 2017 valuation of $1.4 billion made it the first agrifood start-up "unicorn," while Beyond Meat's ascent to become a publicly traded company in 2019 valued at over $1 billion made it the second. Many we interviewed referred to Beyond Meat's public offering as indicating agrifood tech had come of age.

Much of this investment builds on and contributes to the hype that a historically game-changing sector can bring its magic and make much needed transformations in food and agriculture. And the sector comes with big promises. Exclamations of needing to feed the coming ten billion, to meet food's grand challenges of climate change and sustainability, and to develop "moonshot" technologies so radical and transformative as to seem unthinkable are pervasive at the many events and conferences convening organizations sponsor every year, as if saying so will make it so. Whether these claims are sincere is immaterial. They produce the sector's justification for existence, without which investment would not be so robust.

The question is what exact solutions the tech sector could bring to a domain that, despite the hype that it was ripe for disruption, had long been subject to technological innovation, albeit under different institutional configurations. Could Silicon Valley's credo of "moving fast and breaking things" work in the food system in the same way it had in taxis and hotels? If the sector was serious about improving upon chemical-intensive, animal-harming, industrial food production for being unsustainable and inhumane, what could it do differently or better than existing alternatives such as organic, regenerative, or diversified farming, or whole food diets and traditional vegetarianism/veganism? And what could Silicon Valley's strength in engineering and digital technologies bring to food and agriculture? Food, after all, is not code. A quote from one of our interviews captures the entire phenomenon:

> This guy worked at Tesla for seven years and he's done being an engineer at Tesla. He wants to do something meaningful. He wants to go into food. So food is engineering, right? Wrong. But people can tell a great story, they can connect with their buddies who made a lot of money on stock in Tesla [and raise money and] then they go seduce a new VC who's trying to build their own track record in food, who doesn't know anything about food, and they get some more money, and then they get in the paper. That's how it works.

## Objects of Disruption

Understanding the solutions Silicon Valley brings to the table (whether the tech sector can live up to the hype), requires stepping back and considering what it claims to be "disrupting." Recurring references to feeding the ten billion suggests more of the same—that is, increasing food output (see this discussion in chapter 2). Nevertheless, the equally bold appeals to the environmental damage and resource overuse associated with current modes of food production suggest desires to solve problems well beyond hunger. It thus follows that the objects of disruption are the effects or insufficiencies of past technological introductions, many a product of productivism. Since Silicon Valley has stepped to the plate to address these problems, it suggests an additional desire to disrupt the intransigence or ineffectualness of traditional institutions of technological development for food and agriculture.

### *Industrial Food and Farming*

Unlike computing, food and agriculture are deeply entwined with biological processes, such as reproduction, growth, disease, contamination, and decay. Food must maintain its biological integrity from seed to gut to retain its edibility, not to mention to remain delicious and delightful in the ways that attracts eaters, including the coders.[5] The biophysical basis of food and agriculture—comprising soil, microbes, plants, animals, human bodies, climate and seasonality, and much else—have challenged food producers from time immemorial. The vagaries of these elements give rise to all manner of risk: soil depletion, pests, rot, disease, crop failure, and so on. Virtually all food and farming innovation since the

Neolithic revolution, when human food provision largely transformed from hunting and gathering to farming and animal husbandry, has aimed to address these fundamental biophysical challenges.[6]

In their classic book *From Farming to Biotechnology*, coauthors David Goodman, Bernardo Sorj, and John Wilkinson depicted two long-term trends in the development of technology around food and agriculture. The significance of these trends is that they created many of the problems to which Silicon Valley is responding; at the same time, they foreshadowed the kind of solutions being pursued by Silicon Valley.[7] One trend they named "appropriationism" to refer to technologies that could replace and supposedly improve upon processes and inputs once produced on the farm by producing them in factories and selling them back to farmers. This, they argued, would effectively reinforce the rural basis of production by making farms more like industry. Recounting the history of appropriationism, they traced the shifts from mechanical technologies to replace labor, to chemical technologies to increase fertility and control pests, to biological innovation to both enhance yield and control pests. By acting directly on plant and animal biologies, many appropriationist technologies would reduce risk not only from pests and diseases but also from weather, seasonality, and general unpredictability affecting the farm. Appropriationism, in these ways as well, would contribute to farm productivity.

Appropriationism did not necessarily help farmers, however. The heart of appropriationism economically is the commodification of processes once found on the farm to be sold back to farmers.[8] Take farm fertility, for example. A farmer can maintain fertile soil by cover cropping, composting, fallowing, and crop rotations, but those also take space and time away from growing marketable crops. So the farmer buys fertilizer from a company. The use of fertilizer enhances farm productivity not only by stimulating growth but because it allows all the available land to be devoted to crops. That supposedly makes more money for the farmer. However, the farmer has to pay the supplying company for the fertilizer, cutting into profitability. The fertilizer company, in other words, appropriates some farm value. And, as we saw in chapter 2, increasing productivity itself is not generally better for farmer profits.

The other trend Goodman, Sorj, and Wilkinson identified they called "substitutionism." Here the authors referred to the gradual shifting of rural production to factories, made possible by the availability of cheaper or industrially produced raw materials. This, they argued, would at first shift value away from rural production and eventually reduce its importance altogether—also not good for farmers. The shifts they tracked went from preserving (canning, refrigeration) to imitating (margarine) to synthetic substitutes (mainly in textiles and industrial materials, not food) to microengineering food products. Most presciently, they predicted the advent of food fabrication and cellular technologies almost completely attenuated from rural production, technologies that are now the cornerstone of the tech sector's forays into novel foods. Crucially, by acting on the biology of food, substitutionism, like appropriationism, would make food production more controllable and predictable, largely through extending shelf life and minimizing contamination as well as by relying on inputs less subject to biology's irregularities. Substitutionism would also make food cheaper and nominally more appealing. As substitutionist technologies became ever more technical and reductionist, though, they also became less transparent.[9]

To the extent that techies genuinely aim to address the food system's grand challenges, they must address the kind of problems that these past technology introductions have produced. While many farming technologies were sold as solutions to the risk and variability associated with biological production, such solutions have since become problems. Appropriationist technologies have in various combinations contributed to the array of human health, animal welfare, and environmental problems attributed to industrial agriculture. These technologies include extensive use of pesticides, herbicides, synthetic fertilizers, and other deleterious inputs in crop production, as well as confinement, antibiotic and hormone use, reformulated feeds, and simulated reproduction in intensive livestock production. These problems go far beyond the rather well-established diseases of pesticide exposure (e.g., cancer, neurological damage, respiratory damage) to include antibiotic resistance, highly stressed animals, and excessive planet-warming greenhouse gas emissions. The same can be said for substitutionist technologies. Much of food science has been devoted to making food more attractive in taste, texture, and visual appeal as well as less perishable, ostensibly safer, and cheaper. The long list of additives on a brand-name bag of chips is doing just those things. Yet these same innovations have contributed to low nutritional values and in some cases serious health-impairments when additives and processes are toxic, carcinogenic, or metabolism-changing. These problems, too, were born of solutions!

More recent innovations in food and agriculture, largely based in biotechnology, were putative attempts to address some of the problems brought by past innovation. One of the core justifications for genetic engineering in agriculture was that it could replace more injurious chemical or pharmaceutical treatments. For instance, inserting genetic material from a bacterium that is toxic to insect pests (*Bacillus thuringiensis*) into the seed was heralded as a way to reduce the need for chemical sprays. Never mind that this technology, owned by Monsanto and most used in corn and cotton, sped up the timeline by which pests become resistant to treatment. Likewise, biosynthesizing beta-carotene directly into rice grains, the technology involved in Golden Rice, was touted as an answer to vitamin A deficiency, which can arise in part from diets of (nonfortified) highly processed food.

Even more recently, the Enviropig has been genetically engineered to excrete less phosphorous in its poop so to not contribute to dead zones, while sheep are being bred to fart and burp less so to reduce the greenhouse emissions from mass livestock production. The point is that the list is long of technologies that have been developed to try to solve the problems of past technologies. Silicon Valley's entry into food and agriculture says it can provide better solutions to problems that were once solutions to problems that were once solutions to problems . . . And it lays claim to addressing the social problems that have not been helped by past innovation: farmer livelihoods, farmworker wages and working conditions, food insecurity, and worker and consumer exposure to all manner of toxic substances and nutritionally debilitated food.

## *Traditional Institutions of Food and Farming Technology*

Silicon Valley's entry into food and farming also suggests it can improve upon the traditional agriculture and food research institutions that developed many of these past technologies—and failed to address the social problems. In the United States agricultural and food innovation has often come out of the agricultural schools and food science departments of land-grant institutions. The land-grant institutions were a product of the 1862 Morrill Act. The act directed the federal government to grant thirty thousand acres from the public domain to colleges and universities in every state of the union—yes, a settler colonial land grab that ignored Indigenous uses of that land. The sale of that land could be used to establish programs to directly serve the needs of farming, ranching, mining, and other practical professions, yet "without excluding other scientific and classical studies."

Later federal legislation bolstered the research and extension capacities of the agricultural and food science schools that developed and disseminated many of the mechanical, chemical, and biotechnologies referred to earlier. Since these institutions also housed traditional departments of scientific studies, they were well equipped to integrate knowledge of biology, chemistry, and mechanical engineering and apply them to food and farming. In practice that meant these institutions could offer tools to enhance productivity in farming and preservation and consumer appeal in food. Because of an abiding productivist ethos, however, these institutions received little support for addressing the increasingly apparent social and environmental problems of prior innovations.[10]

Indeed, they were particularly ill-suited to address the social issues related to farming and food. For example, California farmers constantly faced labor shortages as a result of their unwillingness to pay good wages. Farm laborers, almost always people of color, worked and lived in grueling conditions for poor wages, because they had no other options. The agricultural sciences had no solutions for labor strife, and the social sciences never gained equal footing.[11] The field of rural sociology, long ago a vital component of the land-grant colleges and universities, with a purpose to inform policy and practice on the sociological concerns of rural life, had become marginal by the mid-twentieth century. Meanwhile, the newer, and more relevant field of sociology of agriculture (born circa 1980), which had more to say about issues of labor and farm structure, was deemed too critical and thus made unwelcome in these same institutions.[12] And so, they could offer no meaningful solutions for addressing food insecurity and the like either.

Agriculture and food technology has hardly been the sole province of the land-grant system. Private agricultural input suppliers, food manufacturers, and product distributors have conducted their own in-house research and extension. Just as often, private companies have worked in conjunction with universities, many not even part of the land-grant system (e.g., Stanford University). For a long time the private sector and the universities hewed to a division of labor between "basic" and "applied" research. University scientists conducted exploratory research, without necessary commercial applications. Once they hit upon something that could be commercialized, however, they transferred it to the private sector, which would bring it to commercial fruition. Since the 1990s, this division of labor has largely eroded. As public support for university research has declined, university scientists have increasingly aligned with corporations to receive research funding, making their research increasingly commercially oriented (see chapter 6).[13] What makes this model of university-corporate innovation different from the Silicon Valley model is that the latter outsources much of the risk of failure to start-ups, even though these are frequently university spin-offs, and expects venture capital to fund it.

As for these oft-derided powerful agribusiness incumbents, they essentially had accrued their capital and power through the processes of appropriationism and substitutionism. Original agribusiness profits, in other words, derived from either selling technologies to farmers in the case of appropriationism or adding value to farm products through food manufacturing and distribution in the case of substitutionism. The value agribusiness appropriated from farmers made selling technologies much more lucrative than farming itself. The solutions inhering in appropriationism and substitutionism therefore were much better financially for the solution-makers than their targets. As agribusiness companies grew, they combined or even split and recombined to acquire the most profitable product lines—and otherwise achieve economies of scale. The gradual erosion of antitrust law since the 1990s left a set of highly consolidated incumbents: the likes of John Deere, Pioneer Hybrid, and BASF on the appropriation end and Cargill, Unilever, Tyson on the subsitutionism end.[14]

These are precisely the kind of sclerotic, entrenched companies that are objects of disruptive innovation, yet as we saw with the Monsanto buyout, these incumbents provide the much sought exits. In pursuing these exits, agrifood start-ups seem to want their cake of disruption and to eat it too. For their part, incumbents have readily engaged the agrifood tech sector; they gain opportunities to outsource their R&D to Silicon Valley start-ups and otherwise hedge their bets on uncertain food futures.[15] In forging these relationships between start-ups, venture capital, and incumbents, Silicon Valley is suggesting this is a better institutional model for making needed solutions.

## Landscape of Solutions

Given the tech sector's implied belief that it can do solutions better, the question then becomes what solutions it is actually bringing to food and agriculture. As a first cut in answering that question we can return to AgFunder and its overview of investment deals to follow the money. As a reminder, AgFunder tracks global deals, which exceed those taking place in or through Silicon Valley. It provides a fine indication of the tech sector's investments in food and agriculture overall and still closely aligns with what runs through Silicon Valley. In AgFunder's 2022 report, based on 2021 activity, it tracked fifteen categories. Of those categories, "eGrocery"—comprising "online stores and marketplaces for sale and delivery of processed and unprocessed ag products to consumers"—by far drew the largest amount of investment, $18.5 billion of that $51.7 billion total, in 343 deals. In dollars, this far outpaced the next category of innovative food, which drew $4.8 billion of investment. However, innovative food—inclusive of cultured meat, plant-based proteins, and novel ingredients—dominated the space in terms of number of deals, with 424.[16]

Sorting AgFunder categories into three larger categories more typically used in market maps of the agrifood tech sector sheds even more light on these investment patterns (table 1). These categories are "ag tech," which are basically appropriationist technologies, "novel foods and ingredients," which are basically substitutionist technologies, and "food tech." Those in the food tech category are generally oriented toward supply chain management, an area of intervention that was not theorized by Goodman, Sorj, and Wilkinson. Three AgFunder categories did not fit in any of those; they are represented in an "other" category. Most significant of those in this category are what AgFunder calls "novel farming systems," inclusive of vertical/indoor farming. Such systems have elements of both appropriationism and substitutionism. They are appropriationist because they create yet more inputs that must be bought that were formerly found on the farm. Most notably, this includes sunshine itself, which in vertical farms is replaced by LED lighting and energy intensive climate controls. They are substitutionist because they involve the shifting of rural production to indoor spaces, generally not managed by actual farmers.

What this analysis illustrates is that the vast majority of investment is going into food tech. This category encompasses technologies that don't even attempt to improve food or farming, either by yesterday's standards of productivity or today's standards of productivity along with better nutrition and environmental outcomes, although some of them claim to reduce food waste. Instead, they streamline existing processes, with a big chunk devoted to marketing and distribution of food and food-related services. It is telling that "eGrocery" is by far the largest category of investment. This basically refers to apps that allow consumers to purchase food online. It clearly reflects the mainstreaming of pandemic buying habits, but how exactly does eGrocery save the world, the putative aim of the tech sector's entrance into food?

Another significant arena of investment are online marketplace platforms for underutilized farm equipment and commercial kitchens (the two categories of "cloud retail infrastructure" and "agribusiness marketplaces"). Basically, these platforms take the path of disruption charted by Uber and Airbnb and apply them to food and agriculture. It's a good question whether there is even a sizable market for "ghost kitchens" or shareable tractors. It certainly could not be equivalent to the market for taxis and hotels. Embedded in the food tech category are also the many companies that use digital solutions to support business activities, providing anything from logistics to personnel management to marketing support. To the extent that the problems they aim to solve are

TABLE 1. Investment deals by category

| Category | Description, per AgFunder | Number of deals | Percentage of deals (%) | Investment ($ in billions) | Percentage of investment (%) |
|---|---|---|---|---|---|
| *Novel foods and ingredients* | | | | | |
| **Innovative food** | Cultured meat, novel ingredients, plant-based proteins | **424** | **13.44** | **4.8** | **9.27** |
| *Ag tech* | | | | | |
| **Ag biotechnology** | On-farm inputs for crop and animal ag, including genetics, microbiome, breeding, animal health | 209 | 6.62 | 2.6 | 5.02 |
| **Farm management software, sensing, and internet of things (IoT)** | Ag data-capturing devices, decision support software, big data analytics | 161 | 5.10 | 1.2 | 2.32 |
| **Farm robotics, mechanization, and equipment** | On-farm machinery, automation, drone manufacturers, grow equipment | 130 | 4.12 | 0.9 | 1.74 |
| **SUBTOTAL** | | **500** | **15.85** | **4.7** | **9.07** |
| *Food tech* | | | | | |
| **Midstream technologies** | Food safety and traceability tech, logistics and transport, processing tech | 382 | 12.11 | 3.8 | 7.34 |
| **eGrocery** | Online stores and marketplaces for sale and delivery of ag products to consumer | 343 | 10.87 | 18.5 | 35.71 |

| Category | Description | | | | |
|---|---|---|---|---|---|
| Cloud retail infrastructure | On-demand enabling tech, ghost kitchens, last mile delivery robots and services | 185 | 5.86 | 4.8 | 9.27 |
| In-store retail and restaurant tech | Robots, 3D food printers, point-of-sale systems, food waste monitoring, IoT | 374 | 11.85 | 4.2 | 8.11 |
| Restaurant marketplaces | Online tech platforms delivering food from a wide range of vendors | 78 | 2.47 | 3.0 | 5.79 |
| Agribusiness marketplaces | Commodities trading platforms, online input procurement, equipment leasing | 110 | 3.49 | 1.3 | 2.51 |
| Online restaurants and meal kits | Start-ups offering culinary meals and sending pre-portioned ingredients to cook at home | 302 | 9.57 | 1.2 | 2.32 |
| Home and cooking tech | Smart kitchen appliances, nutrition technologies, food testing devices | 100 | 3.17 | 0.4 | 0.77 |
| SUBTOTAL | | 1,874 | 59.40 | 37.2 | 71.81 |
| *Other* | | | | | |
| Novel farming systems | Indoor farms, aquaculture, insect and algae production | 117 | 3.71 | 2.3 | 4.44 |
| Bioenergy and biomaterials | Nonfood extraction and processing, feedstock tech, cannabis pharma | 172 | 5.45 | 2.1 | 4.05 |
| Miscellaneous | E.g., fintech for farmers | 68 | 2.16 | 0.7 | 1.35 |
| SUBTOTAL | | 357 | 11.32 | 5.1 | 9.85 |
| TOTAL | | 3,155 | 100.00 | 51.8 | 100.00 |

*Source*: AgFunder, "AgFunder AgriFoodTech Investment Report," 2022, https://agfunder.com/research/2022-agfunder-agrifoodtech-investment-report.

discernable in their websites and pitches, these applications generally revolve on making business systems more efficient. With many companies not even offering solutions unique to food, "efficiency" becomes a stand-in for greater impact, as if efficiency itself is a greater moral good.[17]

What is additionally striking from following the money is that many of the technologies that are associated with ag tech and innovative foods (aka appropriationism and substitutionism) aren't that different than the innovation currently emanating from universities. University researchers are also delivering robotic harvesters and a suite of technologies supporting precision agriculture (discussed in chapter 5). Some universities are also developing alternative proteins (discussed in chapter 4). I've already mentioned MIT's involvement in cellular agriculture: the University of California–Davis happens to have a program too. This may be an indicator of how much Silicon Valley has set the terms by which universities operate (discussed in chapter 6). It also may indicate that Silicon Valley doesn't really have a lock on innovation. The one area in which the universities seem to be ahead of start-ups is in the area of genetic engineering, including gene editing (CRISPR). Whether that is so because universities have a strategic advantage in that area, or because it's such a political hot potato is a question requiring further research.

Finally, it bears recognition that, based on what my team has witnessed at dozens of in-person pitch events and in viewing a range of industry-generated "market maps," a sizable proportion of solutions in the space only tangentially relate to tech. Many of the products we have encountered are consumer packaged goods (CPGs), some of which are probably included in AgFunder's category of "innovative food." It is impossible to know. In this space,

CPGs generally consist of familiar forms of bars, chips, beverages, and bowls, made with quasi-novel ingredients such as cricket powder, quinoa, or moringa or marketed with virtue-signaling labels such as high protein, gluten-free, fair trade, regenerative, organic, and more. Although they are not techno-fixes, and some of them are old news, these products illustrate how the buzzy ferment of the agrifood tech sector has created new opportunities for a great number of otherwise mundane products to be recast as solutions. Their presence, in turn, makes the sector appear bigger than it is, kind of like the hackles on the back of a dog.

In short, a lot of the sector consists of solutions without a significant societal problem (delivery apps) or products without a novel technology (CPGs). It is as if association with the sector itself indicates a will to improve, although in truth it looks like solutionism on steroids, in which the sector itself presents a possibility for mundane invention to be construed as a solution. All that said, investment in the areas of ag tech, innovative food, and novel farming systems is not negligible. As also shown in the table, these areas together are drawing billions of dollars of investment, investments that tell the world that Silicon Valley–style innovation is a force for change.

Chapters 4 and 5 delve deeper into two areas that have received the most public and scholarly attention, in no small part because proposed solutions in these areas appear to address grand challenges. In the vein of substitutionism, alternative protein presents a fix to conventional livestock production. In the vein of appropriationism, digital farming technologies present a fix of myriad sustainability issues associated with crop management. To understand the limits of Silicon Valley to effect significant change in food and agriculture, a closer look at these two paradigmatic areas is essential.

# 4  *Alternative Protein and the Nothing Burger of the Techno-Fix*

Released in 1973, the science fiction police thriller *Soylent Green* depicts an epic crisis taking place in the year 2022, of the kind Malthus predicted would occur long ago. World population has mushroomed, while pollution and climate catastrophe has led to major food shortages. In New York City, where the action takes place, the population has grown to forty million people. Almost all live in the crowded streets where they line up for their rations of water and wafers produced by the Soylent Corporation, which controls most of the world's food supply. Through underground networks, elites are able to obtain desirable food such as steak and a much-prized $150 jar of strawberry jam, but most people haven't eaten anything resembling fresh food in years. They survive instead on plant-based wafers called Soylent Red and Soylent Yellow, likely made from soy and lentils given the portmanteau of "soylent." Of the corporation's products, Soylent Green, allegedly made from ocean plankton, is by far the most prized, being far more flavorful and nutritious. But in 2022, Soylent Green is in short supply, owing to the dying of the ocean. In trying to solve a murder, [SPOILER ALERT] the protagonist eventually finds that the company is using human corpses as an ingredient of Soylent Green.

Rather than being daunted by this cautionary tale of a corporation dominating the food supply with food-like substances made of nontransparent ingredients, amid heartbreaking inequality, the agrifood tech sector acts as if it has been inspired by it. Conducting the bulk of our research through 2022, the year depicted in *Soylent Green*, so-called innovative food had become one of the core areas of investment. In our analysis of the companies that have come through Silicon Valley, we found that the vast majority of innovative foods were protein-forward products that were concoctions of (nontransparent) technoscientific projects. This includes the protein drink Soylent, whose creator obviously has a weird sense of humor. Corroborating the importance and growth of these products, the Good Food Institute reported that the alternative protein sector—consisting of "companies creating sustainable alternatives to conventional animal-based foods, including plant-based meat, egg, and dairy companies; cultivated meat companies; and fermentation companies devoted to alternative proteins"—saw "a stunning $3.1 billion in investments in 2020 . . . investment [that] was three times more than what was raised in 2019 and 4.5 times more than 2018."[1]

The problem to which Silicon Valley claims to be responding differs only in emphasis and outlook to that depicted in *Soylent Green*. Alongside abiding concern with the potential of widespread protein shortages as the population grows larger and more desirous of meat and other animal products, more contemporary concern with the environmental and health consequences of livestock production and consumption underwrite the tech sector's efforts. Believing it can provide tasty and nutritious protein products without animals, its outlook is far more optimistic. Before considering whether Silicon Valley can deliver a less gloomy future, I want to say a little more about the tech sector's focus on protein.

## Why Protein

In focusing on protein, the tech sector is responding to a problem that has been reframed several times over. Protein has a long history of good repute—the one macronutrient appearing indisputably good. Once celebrated for its contributions to vigor, strength, and energy, protein later retained its halo in relation to the vilification of fats and carbohydrates, both of which at different times have been implicated in obesity and chronic diseases.[2] However, the idea that protein might be in short supply emerged in the 1950s and 1960s, when many international development agencies, including the United Nations (UN) Food and Agriculture Organization (FAO), warned of a "world protein gap" threatening the working capacity of entire populations.[3] Between the 1950s and early 1970s, the so-called protein gap led to the mobilization of multilateral agencies, nongovernmental organizations, and the private sector to address what was regarded as a critical shortage of protein for third world countries. Indeed, the "lack of protein was understood to be the defining feature of the Third World food problem."[4]

A few decades later, concern with protein reemerged in new form as concern grew about the ecological footprint of livestock production, in light of the so-called meatification of diets. Which is to say that many predicted a dramatic rise in demand for meat and other animal products due to the growth of the middle class in countries like China and India. Meeting this demand would translate into huge increases in climate-warming anthropogenic greenhouse emissions, the use of fresh water and land for feed, and discharges of malodorous waste.[5] Spreading recognition that the treatment of most livestock was not only inhumane but also detrimental to human and environmental health further amplified this concern. This is because the vast majority of livestock are relegated to high-density contained facilities called concentrated animal feeding operations (CAFOs).

CAFOs provide animals feed that is cheap and convenient but far from their natural diets. And that feed is laced with cocktails of medications and hormones that prevent disease and spur growth and fertility. Prolific use of antibiotics is possibly leading to widespread antibiotic resistance among animals and humans, thus disabling one of the most powerful medical tools to treat serious bacterial infections.[6] Were that not enough, in CAFOs animals are subject to all manner of uncomfortable procedures to enhance their reproduction, extract their milk or eggs, or otherwise maximize their productivity. CAFOs contribute not only to animal misery; they also cause all manner of injury to the workers who support animals' ever more tenuous lives.[7] CAFOs, indeed, are a prime example of the solution causing the problem. CAFOs infinitely improve efficiency for livestock producers, and they also allegedly protect animals from disease. And yet, many of the treatments animals receive in CAFOs are for the harm that CAFOs themselves cause.[8]

Issues with protein, in short, are implicated in several of the grand challenges that supposedly animate Silicon Valley's entry into agriculture and food: global warming, food insecurity rendered as insufficient protein, excessive waste, and general damage to human health and environments. That makes the salience of these alternative protein solutions no small thing: few other products and processes coming out of the agrifood tech sector directly transform food in ways that can appear to meet grand challenges. To be sure, alternative proteins carry a lot of weight for the entire sector in terms of delivering its much ballyhooed impact.

## Alternative Protein as Techno-Fix

Makers of alternative protein are most definitely driven by the will to improve: many are genuinely concerned with livestock production and believe they have a solution that will work for all. In invoking a dubious protein crisis, some also tend toward solutionism—finding a problem for which their innovation can be a solution. Still, of the three impulses undergirding the problem with solutions, alternative protein entrepreneurs are most informed by the techno-fix. This is apparent in the rationales for development of myriad products that are attempts to replace or redefine animal products.

### *Simulacra, or Animal Protein Just Like the Real Thing*

Makers of animal simulacra offer a techno-fix for the animal welfare, environmental, and to some degree health issues associated with conventional livestock or fisheries. They endeavor to fashion meat, fish, eggs, and dairy out of substances that are not animal-based—generally from soy and peas but also fungi, algae, and kelp. (In truth, these substances are *largely* not animal-based since, sorry vegans, many crops are grown in soil treated with fish meal or compost, the latter of which likely once contained animal scraps or egg shells.) Creating products distinct from the far less technological veggie burgers of yore, these producers deploy an array of techniques, including genetic engineering, to make soy- and pea-based burgers smell like, cook like, bleed like, and taste like meat—to replicate the experience of the real thing.[9] The Beyond and Impossible Burgers are no doubt the most well-known simulacra. But burgers are far from the only product. Developers of animal simulacra also make "shrimp" from algae or kelp and milk and eggs from fermented pea protein designed to taste like milk or cook like eggs.

Producers of cellular meat (aka "lab grown," "in vitro," or "cultured" meat) present a techno-fix to animal agriculture as well, except they place more emphasis on the inhumane treatment of animals. Moreover, they do not so much attempt to simulate meat but replicate the real thing, claiming they can produce meat without the animals. The actual technology, developed in the medical field of tissue engineering, involves extracting actual animal cells and rapidly replicating these cells in bioreactors. Replicating an animal cell is one thing; having it have the texture, appearance, and, for that matter, nutrition of a cut of meat is another. Cellular meat companies continue to work on techniques to make what are otherwise blobs of protein more like muscle meat by, for example, scaffolding the cells on structured material. As of 2023, because of this difficulty as well as the costs involved, cellular meat is just coming on the market in the United States, first destined to be sold in high-end restaurants.[10]

What makes the simulacra products quintessential techno-fixes is that developers of these products believe they have the way to address a problem without asking much of consumers. Many developers of these products, themselves highly committed to veganism, assume that everyday consumers do not care enough to stop eating meat, eggs, and milk unless they are eating something functionally and aesthetically equivalent to meat, eggs, or milk. One entrepreneur we interviewed told us he was driven in part by realizing "it was going to be very difficult to get people to compromise and, say, eat the sustainable thing instead of the delicious or convenient thing." Speaking at an event, another producer explained that despite the catastrophic threats planetary life faces due to the use of animals in the food system, "no way are you going to get people to change their diets and stop wanting to consume these foods." In assuming the role of knowing what is best for others, alternative protein entrepreneurs act also upon their will to improve.

Still, their justifications for making simulacra are mostly in keeping with old justifications of the techno-fix: that it is undesirable to change society (e.g., how animals are managed) or too hard to change consumer behavior, and that it is much better to maintain or even improve current lifestyles. Speaking at an event, a representative of a simulacra company said that it was "really about making the most delicious meat, fish, and dairy products the world has ever seen. So that people choose to eat food that is more sustainable than the current conditions, current production system. It shouldn't be a sacrifice; it shouldn't be anything else; it should be 'This is just really tasty stuff.'" In an interview an alternative protein advocate explained that people know about the problems with meat, but they just keep eating more and more of it anyhow, "so rather than continuing to beat our head against this wall of like, trying to educate people out of what they're just going to do, let's change the meat." A speaker at another event we attended said: "It's easier to change the world than it is to change consumer habits."

These products are solutionist too, in that techies construct the problem to be amenable to what they can or want to offer: in this case substitutes for animal products. That takes off the table other solutions, even techno-fixes, which might address the underlying cause of the problem—namely the human practices and political economic exigencies that have made animal agriculture so detrimental. As the geographer Russell Hedberg has noted, it is an approach that problematizes animals' bodies rather than the conditions in which they are raised.[11]

## Protein Ingredients, Definitely Not the Real Thing

Drawing on a distinct rationale from makers of simulacra are those entrepreneurs who aim simply to make more protein, often in the form of ingredients for other foods, if not direct protein supplements. By doing so, they are substituting not for the form (the burger, the egg) but delivering the nutrient itself. This particular techno-fix harkens back to the protein gap era. Following the announcement of the protein gap, the international development community moved "swiftly from defining the problem to engineering the solution."[12] This led to a slew of nutritional fixes that came in the form of "tasteless powders."[13] Today's entrepreneurs are sourcing and reformulating all manner of novel sources of protein—such as insects, brewing remains, algae, fungi, or microbes—from which to develop their own tasteless powders, though thankfully not from humans à la *Soylent Green*. But they are working on ways to extract protein from such substances as black fly larvae, plastic, and even air.

Construing the problem as insufficient protein means any and all ways to make more of it is a fix. Capturing this sense of possibility, a representative of a company experimenting with fermentation said this at an event:

> What I'm really excited about in particular is seeing food 2.0. That's, I think, where companies like us will come in, but ultimately where these more high-tech companies will be coming in, is that it's not just about replacing what's already out there. It's

about how do you use the same Legos but build something new. Create foods that don't exist right now, and create things that are just better across every dimension, that just blow everything out of the water.

Another company representative working with fungi exclaimed:

> What we're trying to do is be part of the solution as we grow from seven billion to nine or ten billion people, how do we produce 70 percent more food? You can't do it through animal-based protein. You've got to have animal-based, but you've got to have lots of other options. We're going to need all the options. We're going to need cultured meat. We're going to need plant-based stuff. We're going to need as many things as we can in order to produce enough protein to feed the world.

These protein ingredients are thus techno-fixes in the vein of the Green Revolution—a technical means to produce more food. Yet they are solutionist as well, in the sense that they came into being through technical experimentation after which developers latched onto the notion of a protein crisis as the problem they are solving. Either way, they are guided by the classic Malthusian assumption that hunger is a product of insufficient production. As it happens, during the protein gap years, those tasteless powders failed to improve the nutritional status of the third world poor because, in fact, making tasteless protein powders available did not address the causes of malnutrition.[14]

Today, there is no further evidence that a protein shortage exists or will in the short- to medium-term future. And if it did, there is no evidence that distributing protein powder would work any better this time. Which is not to say that everyone in the world receives adequate nourishment, but that, following Amartya Sen and his successors (discussed in chapter 2), the problem of malnutrition in modern times has rarely stemmed from insufficient gross production. That alone suggests a problem with this particular solution.

## Assessing the Promise: Alternative Protein as Substitutionism

Given the weight alternative proteins carry for the agrifood tech sector in delivering serious solutions, the ability of these solutions to deliver warrants close attention. It would be challenging to argue that alternative protein products aren't better in terms of animal welfare issues when compared to CAFOs. Using no animals or small biopsies of animal cells is an infinite improvement upon the lives of hundreds of thousands of chickens, pigs, and dairy cows in lifelong containment, eating food to which they are not suited, and being coerced into reproductive practices that suit human timelines.[15] However, few alternative protein companies are making their public case around animal welfare, a result of marketing decisions to bring along flexitarians, those who aren't fully ready to give up animal products.[16]

Instead, most of these companies argue—strenuously—that their products are more sustainable environmentally than animal-based production and nutritionally equivalent or even healthier than animal products. The most dominant environmental claims revolve on comparisons with animal agriculture in regard to resources like land and water, greenhouse gas emissions, and waste, while the health claims largely, but not solely, revolve on nutritional concerns.[17] A maker of simulated egg products, as just one example, once claimed on its website that their process uses "less water, land and energy to achieve equal or better results when compared to current production practices" while providing end products that are more "consistent, reliable and sustainable."

To consider these far more substantial environmental and health promises, I want to return to the concept of substitutionism. Few agrifood technologies hew so closely to the logic of substitutionism: bringing food production into factories in order to change its character. Seeing these continuities sheds light on how alternative proteins function as solutions. Bringing food production indoors wasn't entirely the rationale for substitutionism in previous eras. To make a long story short, at the very beginning, substitutionism was deployed to overcome the frictions of distance and perishability through preservation (e.g., canning, drying).[18] Later, geopolitics became more significant, as countries wanted to find domestic sources for key goods like sweeteners rather than rely on uncertain colonial relations. The substitution of beet sugar for cane sugar, for example, was born of this motive. Then food manufacturers and consumers wanted to cheapen (or sometimes improve) food with more cheaply produced rural products closely analogous to the real thing, leading to, say, the substitute of plant oil–based margarine for butter. Over time, substitutionism became all about marketing manufactured foods, while keeping them cheap and desirable. This involved all sorts of additives to replace or even enhance the goodness of "real food," providing color, flavor, mouthfeel, and even the nutrition that had been lost as a result of all of this processing.[19] Think Flamin' Hot Cheetos, without the nutrition part.

Pursuit of these ends required ever more processing, which meant that more food production (or extraction) moved from farms, ranches, and the sea into labs and factories. That, however, did not eliminate the need for rural production.[20] While some flavoring, colorings, and texturing agents could be made entirely synthetically, food manufacturers still needed the grains, beans, and seeds from which to extract sweeteners, fats, and flour as well as other additives that would emulsify, flavor, preserve, and more.

## *Unpacking the Environmental Claims*

Developers of alternative proteins are operating with somewhat distinct goals from past substitutionism. They want to keep their products cost competitive, but cheapness is not the aim. They definitely want to make their products good to eat—they devote a good deal of effort attempting to replicate the taste of animal proteins or to make those high-protein bars palatable—but that's in service of what they claim is their primary aim: to make their products environmentally better than those derived from animals. Nevertheless, by virtue of the kind of techno-fix these developers are proposing, which revolves on replacing animal products rather than improving how animals live, making them "environmentally better" is tantamount to removing protein production entirely from rural production—or trying to. Think about the problems they want to solve. It is in rural settings where animals feed and fart, land and water become scarce, and waste goes. It is as if bringing production into labs and factories somehow eradicates environmental damage. Consider this claim from a maker of a protein "so pure it is literally born out of thin air": their processes allow them to "completely disconnect from agriculture." In other words, for alternative protein producers "environmental improvement" means embracing the logic of substitutionism.

But it doesn't work like that. Like all substitutionism, factory or lab production does not escape the need for material resources. Something doesn't grow from nothing. For starters, you still need biological inputs either as the base ingredients or to feed the material that is supposed to replicate. While alternative protein entrepreneurs obliquely acknowledge this when they make claims of "plant-based," they routinely obscure from where all those peas, soy, and mung beans will come and how they will be produced, should those simulacra largely replace animal proteins. Take soy beans, a key ingredient in products such as Impossible Burgers and the meal replacement beverage Soylent. Soy was already controversial before it was taken up by food tech due to production practices that involve monocropping, genetically modified seeds, and the widespread use of the herbicide glyphosate. Soy production in the Brazilian Amazon has been particularly destructive, responsible for widespread deforestation and displacement of smallholders as well as unacceptable toxic exposure thanks to the use of glyphosate.[21] For their part, cellular meat developers and makers of those multifarious protein ingredients almost entirely invisibilize the need for inputs. For instance, they'll state that they begin with just a few cells or, per earlier, protein out of "thin air." The developers are pretty elusive about what makes the cells grow or microbes ferment. They are so elusive I have no clue how they do it.

Making these inputs into edible food then takes a lot of processing. Companies frequently refer to the "improved feed conversion ratios" of plant-based food. By that they mean that the amount of feed required for animals to produce the foods humans eat is far less efficient than were humans to eat the crops directly. This argument, first made in the groundbreaking *Diet for a Small Planet,* is compelling.[22] But unlike what the author of that book, Frances

Moore Lappé, advocated—to obtain protein from a diet of relatively whole grains, legumes, and pulses—alternative protein entrepreneurs ask you to eat highly processed derivatives of plants. For instance, the "pea protein isolates" that go into many simulacra are products of a complex process involving extraction at high heat and acidification. It is not as simple as cutting out the middle animal to shorten the path from plants to humans.[23]

The point here is that such processing requires significant resource use, as bioreactors and factories substitute for rural infrastructures. Like greenhouses, these infrastructures utilize all manner of materials (e.g., metals, plastic), energy, and water. While reciting at length the environmental costs of feeding, housing, and disposing the waste of animals, alternative protein advocates are virtually silent about the resources used in building and housing bioreactors, feeding cells, and energizing and temperature-controlling the bioreactors. If reliant on fossil fuels for energy, the energy requirements alone could contribute to greenhouse gas emissions rivaling those of CAFOs. But even the use of solar energy is highly resource-intensive and waste-producing.[24] Were cellular meat, for example, produced at anywhere near the scale of conventional meat, these demands could easily outweigh any benefits, as an increasing number of Life Cycle Assessment (aka "cradle to grave") comparative analyses of cellular meat versus conventional meat are showing.[25]

To give you a more vivid picture, one candid person from a cellular meat company, speaking privately to me at a conference, estimated it would take a bioreactor the size of a blue whale to produce one burger per week for the entire San Francisco population. And San Francisco is a small city, with a population of just over eight hundred thousand. Along the same lines these bioreactors have to be situated in space. While promoters of alternative proteins continue to accentuate the land required for animal production, they make no mention of the space that would necessarily be required to house bioreactors were these production methods to be scaled up. It is not as if the land would be reverted to forests and prairies. I should also note that the feed conversion ratio arguments elide the complex ecological differences between crop and animal production. Animals can be pastured in ecological conditions where crops do not fare well; it is not as if taking the cows away would make land suitable for crops.[26]

Finally, not only must these systems necessarily use resources, they must necessarily produce waste. The waste may not be unsightly and malodorous pools of manure and urine, but that doesn't mean it is nonexistent and doesn't have to go somewhere. The matter of nonexcreted waste warrants a note here as well. Cellular meat developers highlight that growing only the desirable muscle meat eliminates waste. Why, they say, should we waste all those resources growing out a whole animal when we can just grow the desired meat from animal cells? The fact is that industrially produced livestock production rarely produces wasted animal tissue, as animals are transformed from animal to meat; virtually all of the animal "coproducts" go to other uses, without which the livestock industry wouldn't even be profitable.[27] Which isn't to say it's pretty, but it's rarely wasteful. As it happens, one kind of alternative protein company follows a similar logic of conventional meat production but in a way that glows of environmental beneficence. Companies pursuing upcycling not only claim to reduce waste but to use it as their source material, albeit by-products of nonanimal food and beverage processing. It is important to note that processing waste is *never* environmentally innocent. Recycling itself is an industrial process that uses energy, requires nonrecyclable materials, and produces yet more waste.[28] Just as something doesn't grow from nothing, something doesn't leave no remains.

In short, as an environmental project, alternative protein is very much in keeping with an ecomodernist vision that sees nature "out there" in need of protection and thus aims to confine all the unseemly stuff indoors or in highly delimited spaces. But of course it takes all sorts of "nature" to create the source ingredients, infrastructural material and energy to make food indoors, and that activity must be situated somewhere. What this tells us is that the environmental techno-fix of alternative proteins is somewhat of a sleight of hand: it replaces the sites of environmental malfeasance but by no means eradicates them.

## *Unpacking the Health Claims*

While alternative protein advocates stake their claims on environmental improvement, they rarely miss opportunities to make health claims. Typically these companies invoke the equivalence of core nutrients found in animal products, saying their products offer equal or greater levels of the good stuff and reduction of the bad. A simulacra company, for example, states their products deliver "greater or equal protein to animal counterparts, no cholesterol, less saturated fat and no antibiotics or hormones." Producers of cellular meat insist cellular meat is the same or better for human bodies, absent of antibiotics and pathogens, and potentially engineered to be leaner. Skeptics of these claims raise the issue of whether food produced outside its rural habit is nutritionally equivalent. For example, they note that cellular meat cannot contain the vitamins and minerals typically found in meat, such as iron that comes from blood, vitamin B12 that comes from gut bacteria, and omega-3 fatty acids that are taken up by cows eating grass.[29]

Additional reason for skepticism revolves on the logic of substitutionism. When alternative protein advocates focus on nutritional equivalency—bringing to center stage the miracle nutrient protein—they occlude the amount of processing it takes to make their products palatable, often through the addition of sodium, fats, and sweeteners, along with the even more questionable processing additives. Particularly, as they make food from ingredients to which humans are not accustomed, they run the risk of making food that is not digestible or bioavailable as human food. This has been a recurring theme in modern-day substitutionism, best instantiated by products designed to circumvent metabolic processes in the name of weight reduction. Olestra, for example, which provides the mouthfeel of fat without digesting as fat causes all sorts of gastrointestinal disorders.[30] Another concern, totally in keeping with modern-day substitutionism, is that the processes and ingredients that might cause us concern are obscured behind ingredient lists that themselves reference processes to which we have no knowledge. Owing to the highly competitive environments in which these companies operate, they are further obscured behind patents and trade secret protections—a virtual necessity to obtain venture capital funding.

Sit with this for a moment. It is surely striking that the set of technologies coming out of the agrifood tech sector that are most promising for addressing the grand challenges of the food system are not really available for us to understand or scrutinize. As such, another problem with the solution is that we as the public cannot really determine if it's a good one. In effect, we are entrusting those developing and promoting these products to decide for us.

## On the Problem with "Alternative" Protein

This brings us to a remaining aspect of the techno-fix as it applies to alternative protein. As we saw in the Green Revolution, the solution was constructed to be amenable to elite interests, to not rock the boat in other words. You might think that alternative protein does rock the boat, that surely it must be undermining conventional livestock interests. Alas, the facts on the ground do not bear this out.

Despite alleged aims at disruption, many alternative protein start-ups, in their efforts to meet venture capital demands and garner funding, as well as to gain access to expertise and markets, have aligned with the very companies entrenched in conventional livestock production.[31] For their part, many of the big livestock companies that are the supposed targets of disruption have opted not to fight the alternatives but invest in them. Tyson, for example, the world's second-largest processor and marketer of chicken, beef, and pork, was an early investor in Beyond Meat and Memphis Meats, through their venture capital arm. Other traditional agrifood giants including Cargill and Perdue have established their own venture capital division and are developing new plant-based protein products in-house.[32] Repositioning themselves as "protein companies" has allowed these companies to hedge their bets, spread out risks, and otherwise expand into new sectors, while continuing to be heavily vested in conventional livestock production. The possibilities of these alignments for solving the problems for conventional livestock production are not very convincing. Instead, it seems alternative proteins are destined to coexist with conventional animal products to become a market niche.

## Obscuring the Response

In sum, the evidence is light (at best) that alternative proteins are or can be effective solutions to the multiple environmental problems with conventional livestock production, nor is their case for addressing protein deficiency compelling. Their weakness lies with the problem with solutions. They have been forwarded by techies who really do want to make improvements but have framed the problems in ways amenable to what they can bring technologically and can be funded. As such, alternative protein entrepreneurs do not challenge those interests most responsible for the problem. Rather than a response to the problem, based on an examination of what is needed, alternative proteins are a narrowly construed solution to the problem. Just as concerning is that these alternative protein solutions are couched in optimism, but an optimism that actually makes other possible approaches seem unreasonable.

In writing about the problems, really the horrors, with geoengineering as a techno-fix for climate change, journalist and activist Naomi Klein makes an even sharper point.[33] In her review of just some of the ideas that have been put forward, most of which involve reflecting back solar radiation, Klein discusses their potential to change climates in ways far beyond human control and, moreover, in ways that would exacerbate climate injustices irrevocably. What makes these proposals particularly scary, she notes, is that such solutions are not designed to manage a particular problem on earth but the earth itself, the very planet that sustains us. Klein writes that geoengineering is a Plan B, forwarded by those whose financial interests would be harmed with Plan A. But here is the kicker: we haven't even tried Plan A, which is to drastically reduce our use of fossil fuels!

It would be difficult to sustain that the techno-fix of alternative proteins is as potentially catastrophic as geoengineering, but it shares the tendency to obscure, make seem impossible, what might be otherwise. Which is to say that if we're concerned with animal treatment or the environmental impact, there are other responses than alternative proteins. There are many Plan As, in other words, all of which could be responsive. One path would be to encourage and support better livestock practices—those that actually address animal welfare and health directly rather than replace the animals. Pasture-raising livestock, for example, is an existing alternative to conventional livestock production that is much more humane and potentially less environmentally destructive than confinement. Some researchers argue that it actually helps store carbon and replenishes soils.[34] There is a role for technology in such a Plan A, as long as technologies respond to a well-thought-out problem. For instance, a few companies have developed probiotics to strengthen and renew animal reproductive tracts from the damage caused by antibiotic use. This is less a techno-fix than a technology that supports a larger ecological approach to animal health and welfare. One Plan A (response) would make room for more of them.

The core issue is these better approaches make for very expensive animal products compared to those of CAFOs. So that means Plan A has to make CAFO-based animal products less economically viable relative to kinder and gentler approaches. That requires either regulating them much more strenuously, which would both ameliorate some of the worst conditions and increase their costs, or finding ways to financially support the better practices, through subsidies or improved consumer purchasing power. That this particular Plan A seems outlandish politically owes in no small part to the work of the livestock industry as well as a political culture antipathetic to regulation. The livestock industry has fought regulation tooth and nail, even minor measures such as expanding the size of battery cages for chickens or voluntary labels declaring the absence of the use of recombinant bovine growth hormone (rBGH) in milk production.[35] And it also lobbies hard for many of the price supports that make its core inputs, like soy and corn, very cheap.[36] To the extent the livestock industry is more sanguine about alternatives is like supporting geoengineering for climate change, allowing them to continue on as usual, but in this case making room for a nonthreatening alternative that they increasingly support financially. Just as with climate change, you might want to ask why we would want to hand the solutions over to those who caused the problem.[37]

As I write this, it is not at all clear that Plan B, the alternative protein strategy, is working. In 2022, after months of enormous growth, Beyond Meat saw a significant slowing of sales and a concomitant decline in their stock prices, and in 2023 privately held Impossible Foods instigated two rounds of layoffs. Since then, other alternative protein companies have pulled their product lines while retailers and food chains have gotten cold feet in featuring them.[38] In noting this trend, some observers have argued that fast food chains balked at the logistical problems of featuring the plant-based burgers, due to their small volume of demand. Or that the health claims weren't convincing, as many consumers realized how highly processed these products were. A third argument is that the costs of plant-based burgers and nuggets were too high relative to the usual fare.[39] I find this one striking, as it suggests these products faced some of the same problems as those forwarding better livestock practices—that they could not be economically viable without undermining conventional agriculture. This tells us a response cannot avoid doing something about conventional livestock production!

Meanwhile, for those concerned with addressing the problems of animal agriculture more generally, including its climate impacts, there is another Plan A that is not more costly than conventional animal agriculture. This Plan A is in front of our very eyes. It is to reduce or eliminate the consumption of animal products. Diets of pulses, beans, nuts, seeds, and grains provide plenty of protein, as was argued by *Diet for a Small Planet* more than fifty years ago, and eating them closer to the source requires far less resource-intensive processing. No techno-fix necessary. That this Plan A has been forgotten amid the fervor around alternative protein tells you how effective Silicon Valley has been in anointing itself the sole purveyor of solutions—a very gloomy premise indeed.

# 5  Digital Technologies and Plowing Through to the Problem

In 2018, I attended a Future of Agriculture event at one of the innovation hubs in the Boston area. The theme of the evening was data digitalization. The event featured a panel of four, followed by a Q&A session and the requisite networking. Panel members included representatives of a major agribusiness incumbent, an international accelerator aimed at helping to "build great big businesses that can make a difference in whatever industry they are in," a recently formed robotics company founded by an engineer with experience in aerospace and mechanical fields, and a major ag tech company whose founder had come from the field of biology. All the panelists or the companies they represented had come from fields unrelated to agriculture but felt they had something new to bring. One panelist claimed to have not known previously that digitalization was a word. All claimed they had an idea that would make farming more profitable.

I found their comments fascinating. For example, the accelerator representative further described that the company's goal was to "work with really, really awesome founders and kind of fill gaps where they need it." His excitement about working in this space derived from it seeming so, well, cool. To quote him:

> I think one of the cool things about this space, the food space in general, is it affects everyone, everybody in the world. A lot of start-ups don't have that opportunity. . . . One of the cool things about this program and not just this program but innovators and entrepreneurs, people that are working in this space, is that you have the opportunity to build really cool, big things that literally affects every single person on the planet. That is really cool. That was why we got interested in it and why I got interested in this space in the first place was that there is a lot of opportunity to do it. The timing seems right, right now. There is a lot of really cool entrepreneurs like [other panelist's] company and like [other panelist's] company. There are a lot of really neat things happening and it really started—the real focus and explosion of venture capital and interest from entrepreneurs in this space kind of happened with Climate Corp. [that] got acquired seven years ago now. It is really new. Ag tech—that buzzword—is about seven years old and so it is still really new, it is really exciting. It is an old industry that has not had a lot of innovation, has not truly been disrupted. There are neat opportunities to do that. As far as like some of the cool stuff you have seen, it is crazy, it is all over the place. We are taking a bit of a broader approach than just the ag tech. We are thinking more about the entire food system.

He went on to use "cool" many more times in his presentation with very little reference to (much less analysis of) what needs fixing in the food system. Moreover, in suggesting his company could build any great business, he displayed little understanding or concern about food systems, farming, and the particular problems they pose—which are not generally all that cool.

Just as striking to me were the comments of the robotics company representative who was working on "a completely automated drone solution for agriculture." As he put it:

> One of the biggest problems that most farmers face is not knowing which part of their fields may need to be [treated]. . . . They know that there is a problem in the field and they spray the entire field. That solves their problem but it is not the best way to solve the problem. All those chemicals, all that stuff is going to go somewhere and end up in our bodies at the end of the day. . . . This is where the concept of precision agriculture started, where you try to address a much smaller part of your fields.

He explained how the original technologies like satellites and the Landsat program didn't provide adequate resolution for precision agriculture. Enter consumer drones, which were not adequate for farmers because it takes several days to image the data, making "the whole concept of precision agriculture . . . not really as precise as it could be." The solution, he averred, was to "take out all of the manual flying associated with drones." He continued:

> It is not just a matter of flying, it is also you need to know where your field boundaries are, you need to fly the drone, you need to transfer the data, you need to process the data, you need to access the data, you need to understand the data. That is a lot of time for a farmer or an agronomist to deal with when they have a lot of other things that are much more critical for their livelihood. Auto-

mating that entire chain, that workflow, is an easy path for farmers and agronomists to get access to the [necessary] data both in space and time. . . . We can now have the ability to get data essentially on a daily basis. You have a pest infestation that is beginning in your farm, you don't need to dust your entire farm. You take that corner or piece of farm where the infestation is starting and you address it on day one . . . And that is just even without putting any smarts on the data. . . . Like all kinds of machine-learning algorithms, whatever other buzzwords that entail.

His point, as I saw it, was that ever more precise and easy ways of gathering farm data would give farmers better ways of managing the farm in ways that would reduce the use of undesirable inputs.

The rationales I heard that evening were quite distinct from those I would later hear at events or panels focused on alternative proteins (discussed in chapter 4), in which speakers passionately described how the myriad problems in animal agriculture led them to want to provide an alternative. What I witnessed that evening was less familiarity with the problems of agriculture other than it seemed "cool," but high aspirations that applying technologies borrowed from other sectors to agriculture could both reduce input use and restore farmer profitability. While alternative protein, in other words, came out of a definable set of problems to which innovators could pursue a techno-fix, the opposite could be said of digitalization, which was born of available solutions looking for new problems to solve in a characteristically solutionist way.

As discussed in chapter 3, ag tech, comprising both digital technologies and ag biotechnology, represents a somewhat small percentage of worldwide investment deals and volume in the agrifood tech sector relative to supply chain technologies. Yet along with innovative food, the vast majority of which involves alternative protein, digital technologies loom large in Silicon Valley imaginations, garnering near equal hype in terms of being world-changing, if not exactly world-saving. (No one is claiming that eGrocery will address the grand challenges of the food system, for instance.) Along with alternative protein, these technologies have received the most scholarly attention, too, precisely because of these promises. Nevertheless, more than emanating from different impulses, they approach the grand challenges of food in very different ways. While alternative protein continues a long legacy of substitutionism that brings food production indoors, digital farm technologies continue a long legacy of appropriationism that reinforces rural production.

As with alternative protein, understanding how these technologies continue a prior tendency will help us answer the critical question of whether they bring improvements to past ways of farming: in other words, whether they are good solutions. However, it is important to establish how digital technologies became a considerable solution to farming in the first place.

## The Coming of Digital

We should not be entirely surprised that so many Silicon Valley interventions are digital. Notwithstanding its geographic proximity and cultural affinities with the life sciences industry, innovation in Silicon Valley has long centered on computing and communications, and more recently branched into big data, artificial intelligence (AI), internet of things (IoT), and other realms of digitalization. As discussed in chapter 3, many entrepreneurs left coding and the traditional tech sector of Silicon Valley to work with food, attracted by its materiality, its pleasures, its potential for impact.

Other centers of agrifood tech, like Boston, have followed similar trajectories. Tracing how these technologies were transferred to food and agriculture and for what possible reasons sheds light on the solutionist underpinnings of digital farming technologies. The migration of the digital into a domain so essentially biological began with precision agriculture, and in some ways so did the tech sector's entry into agrifood. Precision agriculture technologies, referring to a range of digital tools such as sensors and drones that can monitor field specific conditions and provide information, first came on the scene in the early 1990s. Precision agriculture arrived at a time of decreased support for public agricultural research and extension, and private sector firms first introduced some of these technologies. With agricultural research and development (R&D) long guided by a "supply-side" mentality, in which scientists, scientific institutions, and technology developers have habitually promoted technologies as solutions regardless of stated need, it didn't much matter that farmers had never asked for a solution like precision agriculture.[1]

Rather, a search for new uses for technologies that had been developed in nonagricultural sectors such as manufacturing and the military led its promoters to farming.[2] The same could be said for the "big data" analytics that came to wow the tech sector in the 2000s, especially as digital tools saturated more profitable markets in health care and finance.[3] Once agrifood tech became a thing, though, the sector embraced these digital agriculture technologies—and digital agriculture entrepreneurs—as its own. All of a sudden these technologies became radical and disruptive, even "game-changing" enough to "shore up continued investment" in the tech sector and give rise to "a host of startups with minimal previous connections to agri-food."[4]

## Digital Agriculture as Solutionism

But what exactly is the big problem that digital agriculture technologies could solve? Typical rationales are nearly identical to the objectives expressed at the aforementioned event I attended: to improve farmer yields and reduce the use of environmentally destructive farm inputs, both of which promoters claim will make farming more profitable.[5]

The thing is, if you first identified insufficient farm productivity or excessive resource use as a problem, you wouldn't jump to better visualization and data processing as the obvious solutions. Likely you would turn to the many technologies that have already been tried—from plant genetics and fertility regimes to improve yield to all manner of water saving and pesticide reduction programs—to lessen your environmental footprint. Here it is important to emphasize just how different precision agriculture technologies are in farming than these other methods. The value proposition of precision agriculture is that more data about the complex, unwieldy, and risky ecosystem factors inherent to food provisioning can improve management of biophysical systems. Sensors, drones, and information software may assist farmers in better visualizing what is happening in their fields to inform their decision-making about what inputs to use. But these digital systems do not treat the fields directly or suggest inputs less damaging to biological systems than past technologies.

Given that these technologies could at best provide indirect support to managing fields, landing on more information as a solution was a textbook example of solutionism in which those with an already conceived technological solution found a problem for which it could be put to use. The extension of digital technologies into the larger domain of food and agriculture, including supply chain management discussed in chapter 3, further indicates the solution-driven nature of digital technologies.[6]

## Demonstrating Solutionism

Having glimpsed such solutionism before my research officially began, both at the Boston event and the FoodBytes! event described in the preface, one of the research questions that guided this project was how entrepreneurs land on their solutions. Almost needless to say, in answer to our interview questions about how they arrived at their solution, virtually no one described a deliberative process by which they first defined a problem and considered available solutions. Mainly they wanted to find a way to apply tech to a problem.

Still, in contrast to the alternative protein entrepreneurs who could cite chapter and verse of the problems with livestock (or protein shortages), many digitally-focused interviewees described how they entered the sector, often straight out of tech, with ideas of problems they might solve. Some of these more circuitous entry stories among digitally-oriented entrepreneurs show the degree to which they put the solution cart before the problem horse, some to the point of never even articulating a problem.

The founding story for an exchange platform typifies this searching-for-a-problem dynamic in stark terms. Two generations removed from farming, one founder had always had an affinity for agriculture, but his academic training in neuroscience funneled him into Silicon Valley along with "90 percent of neuroscience PhDs." After working in a few different health tech start-ups, he happened into ag tech through a design event where he met his cofounder. From there they "pretty quickly just sort of latched onto the shared sense of how fundamentally the main problem was to just make farming a desirable thing to do. And so we looked at commercial agriculture through that lens and saw just really similar solutions." And yet, despite their alignment with each other, they never quite landed on a solution aligned to the sector. They began as an equipment-sharing platform, in their words an "Airbnb for tractors," but then came to the conclusion that farmers needed help negotiating with big agricultural dealers—"the people that sell things to farmers, like their chemicals and machines." So they decided to shift away from the equipment rental business and become middle players, of sorts, to negotiate better input prices for farmers. Involving more phone calling than anything else, their technological aspirations fell by the wayside. Within a couple of years they were out of business.

Another company whose founders claimed passion about empowering farmers parlayed expertise developed from decades in finance to deploy blockchain and then IoT technology to food. As the manager we interviewed put it, they "got tired of the fintech space [referring to the application of technology to financial services], and so they were looking for another industry. . . . And so they looked at music and insurance and education and identity, and then landed in the food supply chain." Beginning by "digitizing tomatoes" (whatever that means), they expanded into other food items with the objective to create "a digital twin of the supply chain." By capturing that data, they felt they could support companies in a quest to tell the story of the food they grow and bolster brand integrity. "So if they say that they're sourcing from sustainable farmers or from local farmers," the manager explained, "we're going to capture data that supports that assertion." Latching onto the idea that revealing the conditions of production would be transformative, they never explained how this transparency would actually transform how food is produced.

More closely aligned with ag tech per se was the case of the person who had spent three years building a company that used machine learning to inspect road surfaces. The founder decided "it was a no-brainer" to try such deep learning to look at plants—"it's just taking pictures of plants, not roads"—and thereafter to use the expertise of botanists, horticulturalists, and viticulturalist "to understand the patterns on plants." The company could then "embed that in the AI" and "just sweep across the plants." The founder hoped to sell these "passive data collection platforms" to some "really big tractor vehicle manufacturers" and was "absolutely certain that within five years," it would "be routine in specialty crops or something." When we asked what value that actually brought to growers, he first deemed it a great question and then went on to say that the technology could be used as a doctor or coach to growers. As a doctor, it could look for disease infestation, things that are damaging to a plant and then alert the grower to deal with it, like "You've got a COVID-like disease on vine 6, row 7." As a coach, the technology could suggest "some sort of remedy"—maybe "de-leaf" or "shoot thinning" or "some sort of water issue." Here the founder never quite articulated a problem that such a technology might solve.

In another case in which the entrepreneur pivoted from elsewhere, he eventually discovered a problem that he could solve but it didn't start that way. This person was trained as a mechanical engineer, had a brief stint in investment banking, and returned to engineering to develop computer visioning for robots. One robot the former company helped develop used computer vision to detect dog poop and automation to pick it up in people's backyards.

(No, I'm not kidding.) Yet he realized that "trying to validate consumer demands for the next Roomba, or the next widget is a really tough sell," and in consultation with investors, he decided that addressing "a genuine pain point" was necessary. The company first looked into operations and logistics in developing markets as well as point-of-sale services. But it was through meeting growers as a part of a California-based accelerator program that they learned of the state's historical dependence on manual labor and current farm labor shortages. "That's how [they] came into ag tech. . . . It's probably the most impactful way you can get technology out the door, which is to solve a problem, which the world is currently really challenged by." Eventually they settled on a weed cutting technology for leafy greens. Although it remains widely controversial whether California's historical farm labor issues are appropriately solved by robots, in this case the company did eventually land on a problem for their solution, notably through attuning to potential users.

That solutions sometimes beg the problems in the sector has not been lost on observers of the sector, including consultants, investors, and entrepreneurs themselves. While all who spoke to this issue noted the existence of companies genuinely interested in solving problems, we heard many cross-accusations of solutionism. "Unfortunately, there's a lot of snow jobs that are going on right now. It's like, 'Hey, we're going to go change the world,'" one consultant and investor commented. Another industry expert and investor put it even more bluntly: "So they're not actually fixing a problem. They can't." He continued:

> I know people who have raised $5 million without ever talking to a farmer. . . . Because they watch the news that farmers have a prob-

lem with drought, and water, and nitrates in the soil, and all this other. . . . "Oh, well, I can fix that." Four years later, they haven't fixed it. They've pivoted to do something else that makes sense to their investors to try to get some money coming in. So people don't listen to what farmers need.

An employee at a start-up firm captured the issue even more succinctly, describing this common entrepreneurial approach as the "I found a solution. Is there a problem?" path.

## Assessing the Promise: Digital Agriculture as Appropriationism

It may not seem a big deal if entrepreneurs come late to a problem if indeed what they're selling otherwise makes sense. At first glance a means to visualize fields and process data in order to encourage farmers to reduce use of toxic pesticides and minimize use of precious resources like water makes some sense. It may make less sense when viewed through the lens of appropriationism. For what they're selling is another iteration of appropriationism, albeit with an environmental and philanthropic twist. As discussed in chapter 3, past appropriationism has generally enhanced overall farm productivity by supplying inputs that can help farmers with fertility, pest control, labor saving, and plant and animal productivity. At the same time, use of many of these inputs caused the environmental damage to which tech solutions supposedly respond.

How, then, do digital ag technologies instantiate appropriation? The answer is that the analyzed data that suppliers sell originally comes from farms and farmers. To put it less obliquely, when a farmer purchases a digital technology, whether outright or through a license, they are basically purchasing data compiled from field sensors, drones, and other devices, some attached to farm machinery, which collect data at the farm level. These data can be about the conditions of soil, water, crops, weather, and climate, or the equipment itself. Although some technologies use built-in computers to process that information and deliver it right back to farmers, in the case of ever more present big data, local computers and cellular transmitters upload these data to the cloud to be amassed with the data from hundreds of thousands and even millions of other farmers. Sophistical algorithms and machine learning analyze these enormous volumes of data, far beyond the capacities of humans. The conceit of "big" data is that the volume of data provides exceptionally useful information.[7]

The catch is that rights to those data rest with the companies making and selling those devices, which, like Facebook, collect massive amounts of data from users whether they like it or not. The farmers get access to those data, as potentially useful analytics, through a license, even though some of the data may have come from their own farm. It's a solution that effectively replaces farmers' own observations and firsthand knowledge of their fields with a machine's under the presumption that more data is better data.[8] This solution therefore gives little thought to the investments farmers have in their skills of observation, much less their willingness to hand over their data to these companies.[9] While those selling these technologies portray an upgrading of farming, such that the farmer will sit in their office with their iPad and make decisions without ever touching the dirt, critics suggest otherwise: that it will amount to the de-skilling of farming with decisions basically handed over to artificial intelligence—and essentially the values and ideas of whoever made the code.[10]

As a solution, one of digitalization's promises is no different than virtually every other appropriationist technology that has ever been offered: that it can improve yield and in doing so make farming more profitable. Putting aside for a moment that information delivered to farmers has little chance of itself making plants grow faster, bigger, or more productively, the emphasis on yield is problematic enough. Whether put in neo-Malthusian terms of feeding the ten billion, a ubiquitous refrain at pitch events, or put in farmer friendly terms of increasing profitability, as at the event in Boston, it is wrong. As discussed in chapter 2, hunger is rarely a product of insufficient output. Nor do yield-enhancing technologies help farmers, except for early adopters. In classic treadmill fashion, as increasingly more farmers adopt productivity-enhancing technologies, they all see more yield, creating gluts in the market. Prices decline thereafter, and farmers are back to square one. But here's the real rub: those selling these technologies want to make a profit too, and they do so by appropriating farmers' profits (thus the terminology of "appropriationism"). It remains to be seen how companies selling precision agriculture technologies can both be profitable themselves and bring profitability to a domain that has historically been unprofitable in large part because of appropriationism!

This brings me to the second promise: that of environmental improvement. This objective assumes that better information will encourage farmers to target their treatments to hotspots only, thereby reducing the use of precious resources and toxic inputs.[11] Heed the promise: satellites and data analytics will help reduce use of toxic chemicals, with no mention of less toxic treatments to substitute for the chemical. To be super clear about this, precision agriculture technologies do not provide farmers with alternative treatments. Rather, they visualize, diagnose, and inform decision-making on how to apply existing treatments. And according to recent research, apparently they don't do that very well.[12] That leaves open the question of whether precision agriculture ever reduced reliance on biological alteration and chemical inputs. Hard to say, but scholars have argued otherwise—that precision farming actually has legitimated chemical-based agriculture.[13] You can see that in the claim made at the Boston event: you just try to treat a smaller part of the field—yet with the same chemical!

Another concerning environmental aspect is that digital technologies tend to best support spatially extensive monocultures of "commodity crops" such as wheat, corn, rice, and soy. This is also evident in the comments from the robotics company representative at that Boston event. He alluded to the need to survey thousands of acres, too many for a farmer to manage with a hand-held machine, thereby justifying a robotic drone. And why would a farmer with much fewer acres invest in such a technology? Just as there is little point in investing in a $300,000 combine to manage a five-acre field, there is little need for highly technified surveillance technologies if a farmer can walk the field in a matter of minutes. To the extent that digital technologies effectively encourage monocultures is not a road to more ecological farm management. Monocultures are simplifications of biological systems; they lack ecological resilience and tend to require more chemicals to manage pests. The primary reason farmers grow crops at such extensive scales is because profit margins are so low it is the only way they can stay in business.

In short, digital technologies tend to reinforce the worst aspects of appropriationism, extensive production in order to remain viable and continued use of synthetic fertilizers and pesticides to manage these simplified biological systems. They offer no obvious path to actually improving farmer profitability. If indeed farmers are willing to pay for these technologies, it looks a lot more like a solution for the profitability of companies delivering precision farming technologies than a pathway toward agricultural sustainability. It thus *is* significant that these technologies came from elsewhere looking for a solution. They not only share with other solutionist technologies a raison d'être revolving on lack of information as cause for poor health or environmental ills (think consumer apps to monitor food intake or climate footprints). By reaching "for the answer before the questions have been fully asked," the ensuing mismatch of problem and solution leaves the real problem without remedy: here the intense use of environmental- and health-debilitating inputs in order to grow food.[14]

## Toward Response

As a reminder, Silicon Valley ventured into food and agriculture as a domain in which it could have impact—and even bring moonshot technologies to address some of the most daunting challenges of food and agriculture. I entered the research reported within this book to assess whether Silicon Valley–style innovation was capable of improving on earlier rounds of agriculture and food innovation, by bringing novel agricultural technologies or reformulated foods that are allegedly better for crop, animal, soil, and human health and the biophysical environment writ large. Conceding in advance that most anything Silicon Valley would offer would be a technofix, I was nonetheless surprised at its limited menu.

To recapitulate, a great deal of activity is in the category of food tech, technologies that don't even attempt to improve food or farming, either by yesterday's standards of productivity or today's standards of better nutrition and environmental outcomes. Instead, they streamline existing processes, with a big chunk devoted to marketing and distribution of food and food-related services. Yet the two more promising areas also fall short. As a major subset of novel foods, alternative proteins create an alternative to conventional livestock but do not provide tools or pathways to improve conventional livestock production, much less really undermine it. Digital farming technologies, a significant subset of ag tech, provide ways of visualizing and analyzing how to use farm inputs but leave it to the farmer to choose what inputs to use, which can often be more of the same. And both approaches reinforce the political economic effects of technology introduction: shifting value away from the farm and into the pockets of incumbent input suppliers, crop and animal processors, and marketers. These are not good solutions!

A primary reason that these digital solutions in particular are not well matched to the major problems of food and agriculture is that they have been transferred from elsewhere, by entrepreneurs, trained in a Silicon Valley mind-set, whose entry into agrifood tech was not necessarily geared toward solving a particular problem. Rather, some of these entrepreneurs appeared in the space as a means of gaining access to the capital and markets generated by this much-hyped sector or simply because food appeared attractive or impactful. In essence, they came with solutions looking for problems, and those problems they encountered were not the sort that could be directly solved by the tools they brought. And so they backed into problem statements to support the solutions they could offer, solutions that were more of the same. As with alternative proteins, it isn't clear that their investments in ag tech are paying off, either. Research has shown that farmers are reluctant to cooperate with digital companies to provide data.[15] Meanwhile, adoption of on-farm tech remains low, which some observers attribute to the fact that farmers are users, not beneficiaries, of these systems: they have to bear the costs of new hardware and changed practices and get little from that.[16] This suggests a fatigue with processes of appropriation that reduce farmer viability rather than enhancing it. But it also suggests a realization that these technologies do not address actual problems for farmers—or the environmental problems that stem from farming.

So what would be different in terms of addressing the on-farm environmental problems stemming from past technological introductions? My preferred approach would not begin with the presumption of a technological fix, but rather a coming to terms with the harms of past technological introductions: mechanical, petrochemical, and biological. That would include not only an assessment of how that environmental damage could be reversed but also a recognition of the social harm (and gain) of those past technologies and how they might be rectified. In other words, a response would consider the social-ecological project first—the kind of social and environmental conditions to create, with whom and for whom. That would require attention to the distribution of land, labor, capital, and knowledge as well as the health and environmental harms of farming and pursuing practices and policies that address these.

Such an approach does not make technology irrelevant. Better practices might emerge from the availability of treatments less harsh than past treatments, even some currently being developed, such as those that can control pests with fewer toxic biological substances or biodegradable materials, strengthen animal health and address the damage caused by antibiotic use with probiotics, or create biofertilizer with fermented food waste. Such treatments are more probiotic than antibiotic in the sense of enhancing the health of the desired organisms rather than killing all that threatens them, which tends to cause a lot of collateral damage and harm.[17] This is not an unequivocal endorsement—some of these technologies are still commodities for which farmers have to pay (appropriationism!). But at least they offer a veritable improvement, in no small part because they can support diversified farming, itself an arguably better way to farm.

The irony here is that many of these gentler treatments—we can even call them solutions—already exist or are being developed outside of the tech sector. Farmers and supporting researchers involved in such systems as organic, biodynamic, integrated pest management, and other alternatives have long been trying to develop and disseminate nonchemical means to deal with pests and diseases. However, they have never received the same kind of financial (or ideological) support as the solutions coming out of the land-grant universities or today's tech sector. This suggests that the problems of agriculture and food may be better addressed by those social movements and practitioners who have been in it for the long haul, who have studied the problem, developed treatments, and whose attention to the biophysical realities of food and farming is something to learn from. It may be, in other words, that the problem is not a lack of technology but a lack of funding and research support for those already engaged in developing alternatives to "big ag" or "big food."

It happens that I work at a university that has long pursued this approach, the first in the nation to develop and sustain a program in agroecology. But, alas, Silicon Valley fever has infected there too, threatening to erode the program's philosophical and practical grounding. More broadly, the entire Silicon Valley model—that claims to value impact yet often construes impact as disruption and enhanced efficiency, that requires a marketable solution rather than a holistic approach grounded in deep understanding of the problem, and that encourages competition rather than collaboration to create the best ideas—is being imitated in an institution whose values were once quite different than Silicon Valley's, if not entirely anathema to them. That institution is the university at large. How and to what ends universities have become incubators of a Silicon Valley approach is the topic of chapters 6 and 7.

# 6 Silicon Valley Thinking Comes to the University

Sometime around 2020, I got wind that my campus was planning to develop programs in the area of ag tech. Frankly I was stunned by this news. I could not understand why the University of California–Santa Cruz (UCSC), of all places, would be pursuing this direction, a sentiment shared by many others who study food and agricultural change at this campus. Several aspects of the campus's history made an ag tech initiative strange. For one, UCSC began as an experimental liberal arts institution. The original physical design of the campus, for instance, created colleges with unique architecture and intellectual foci rather than disciplines. Students learned in small seminar-style classes and received narrative evaluations instead of grades. At one point after its founding, many considered UCSC the top public liberal arts college in the country. So a highly technical approach didn't seem to square with this history.

In addition, UCSC was not a land-grant institution, established by the 1862 Morrill Act. It was Berkeley that contained California's original land-grant college, although many of the land-grant activities eventually devolved to UC Davis and UC Riverside. Not part of the land-grant system, UCSC never had agronomics or agricultural engineering departments. Nor did it automatically receive federal funds for conducting agricultural research, although the university can and does apply for USDA funds. Most important, UCSC had developed a reputation for programming in food and agriculture at odds with the goals of much of ag tech. In fact, UCSC was the first college or university in the country, dating back to 1967, to create a farm and garden dedicated to agroecological farming practice.

As a practice, agroecology centers on place-based, on-farm biodiversity in support of soil health and nutrition as well as weed, disease, and pest management.[1] Agroecology embraces the complexity of biological systems rather than attempts to simplify them. Yet agroecology has come to connote more than a practice or even a science to support that practice. As many peasant communities and Indigenous people have long farmed in accordance with agroecological principles, agroecology has come to be one of the pillars of, if not identical to, various social movements by and for small or economically marginal farmers and food sovereignty.[2] Practitioners of agroecology categorically reject the idea that agricultural systems can be divorced from these social and political aims.

Through its world-famous apprenticeship program UCSC had trained hundreds, if not thousands, of people who had become agroecology-oriented farmers, many certified organic, as well as dozens of sustainable food activists. Those trained in agroecology generally spurned productivist agriculture given its propensity to rely on the latest technologies and reinforce large-scale monocultures. UCSC faculty strength in food and agriculture were in fields equally anathema to productivism. These faculty include the many natural scientists who study soils, plants, and biodiversity in support of agroecology as well as a long lineage of social science faculty like myself, who through their internationally-recognized research have in various ways critiqued the role of much technology in agriculture for its adverse social and ecological consequences.

Why would the administration pursue this initiative in a way that could jeopardize its irrefutable impact and undermine its historical reputation? The apparent motivations for the initiative were multiple. Engineers on campus, none with a background in agriculture, saw the potential for partnerships with private sector actors to help fund and disseminate further research. Demonstrating their inclination toward solutionism, they were excited to think of new uses for gadgets they were designing. One engineer even talked of a sensor to be used to help plants detect the direction of the sun, an idea that drew some lighthearted derision from the agroecologists who were pretty sure plants have no issues finding the sun. Engineers were responding to growing interest among students to use engineering for good—in other words, to produce techno-fixes via a will to improve. Finally, in 1994 the university had acquired property from a former military site closer to the southern end of Monterey Bay than the main campus. The idea was to develop the site into a science and technology park, again in hopes of securing additional private funding for the campus at large.

The geographic proximity of UCSC to both Silicon Valley and the Salinas Valley also factored large into these aspirations. Having already developed a satellite campus in Silicon Valley, the university saw the potential for additional partnerships as a means to garner more funding. Its interest in the Salinas Valley was more curious. Notably, it is one of the world's centers of high-value, high-tech vegetable production, really one of the first sites to which the term "industrial agriculture" was applied—coincidentally by a former UC faculty member, William ("Bill") Friedland, whose work was formative in critical agrarian studies. The common agricultural practices in the Salinas Valley are precisely the kind that agroecologists critique. Nevertheless, in internally justifying the initiative, campus administrators suggested it could leverage the campus's long-term investments in alternative agriculture and a campus culture valuing social justice to create a *unique* brand of ag tech compatible with agroecology and the support of small farmers and farmworkers.

The significance of this initiative is not just because it took hold on this particular campus, with its unconventional legacies. To the contrary, it presents a model case of neoliberal logics and Silicon Valley style taking hold of a public university. Of particular note are (1) the imperative of finding external sources of funding; (2) the elevation of applied STEM fields (science, technology, engineering, and mathematics) relative to other areas of expertise; (3) the increased emphasis on technical problem-solving for broader economic benefit; and (4) attentiveness to branding. All are hallmarks of the neoliberal university, which provides another important context and site for contemporary solution-making in the vein of Silicon Valley. Separately, the initiative illustrates the appropriation of ideas founded in agroecology (such as sustainability) and repurposed to justify the same kind of agricultural (non)solutions stemming from Silicon Valley, especially those based in efficiency. As I mentioned in the introduction, such programs integrating tech and food are proliferating on college campuses.

## Making of the Neoliberal University

To understand the university's turn to solution-making, we have to return to the phenomenon of neoliberalism. Universities were once relatively immune from economic pressures, funded largely by state budgets in the case of public universities and tuition and private endowments in the case of private universities. That doesn't mean they were completely separate from the private sector. The very purpose of the land-grant university was to serve farmers and ranchers with the latest technologies, ostensibly to maintain their viability. And some universities have always had close ties with the private sector, even sharing a revolving door. Stanford University's earlier coordination with Silicon Valley was not incidental. The Stanford Research Institute, for example, founded in 1946, was populated by faculty pursuing "science for practical purposes" and "assisting West Coast businesses." These alliances were forged, even at the time, in ways that "might not be fully compatible internally with the traditional roles of the university."[3] Early on, Stanford created a Stanford Industrial Affiliates Program, which for a modest fee granted companies access to academic labs, research meetings, students, faculty, and special recruiting events. In 1969 the university even established a licensing office, which helped commercialize the new inventions that came out of its programs.[4] MIT, whose very brand rests on innovation and engineering, followed a similar path. But the very life of universities in general, in terms of funding and legitimacy, didn't depend on these alliances.

The neoliberal policies and ideas that gave rise to Silicon Valley transformed both the means and ends of universities. To show how a once great public university system succumbed to these pressures, I focus on the university system of the state of California, a story I know well. But it is not an exceptional one. The general tendencies I describe have been well noted throughout academia, both in the United States and elsewhere. In 1960, under the leadership of Governor Pat Brown and UC president Clark Kerr, the state of California adopted a Master Plan for Higher Education, the goals of which were to broaden access to higher education and to differentiate functions among the University of California, the state colleges, and community colleges. Thanks to the wealth generated in California during the postwar period and relatively high levels of income and property taxes on that wealth, the state was able to greatly expand the number of campuses while keeping tuition low for the university and nonexistent in the community colleges. Ample state support, coupled with a robust system of shared governance between faculty and administrators, allowed the university to develop a well-rounded liberal arts curriculum in which sciences, social sciences, humanities, and arts were on equal footing.

To see how these glory days ended, we need to return to Ronald Reagan, who before being elected president of the United States had served as governor of California, following the heyday of Pat Brown (Jerry Brown's father, if you're following along). When campaigning for the governorship in 1966, Reagan used the platform to cast aspersions on the University of California. He detested academic freedom and student protests. Railing on the free speech movement, Reagan famously promised to "clean up the mess at Berkeley"—and the rest of the UC system. During Reagan's governorship and thereafter, the university underwent significant transformation, owing much to the encroaching neoliberal climate. Like elsewhere, California had seen offshoring of manufacturing jobs and high inflation. But California had an additional dynamic, typical of the state: a red-hot real estate market.

As prices on homes and commercial property climbed skyward in the 1970s, state residents on fixed incomes (receiving, for example, Social Security, pensions, or veteran's benefits) suffered. With property taxes based on the value of homes and commercial property, rather than their sales, California homeowners had to pay ever increasing taxes, even though their incomes remained stagnant. And like elsewhere in the United States, older white voters were particularly unhappy to see their taxes go to programs that didn't directly benefit them. Unlike the New Deal, which established Social Security and unemployment insurance for all (although the programs largely benefitted white people), the Great Society programs targeted the Black "inner city" as it was imagined before gentrifiers moved into city cores. A tax revolt ensued, led by two rabid anti-tax, antigovernment influencers (Howard Jarvis and Paul Gann). The result was the 1978 passage of the statewide initiative Proposition 13, which rolled back property tax rates to 1975–76 levels and restricted further increases unless a property sold.

At first the California legislature found ways to paper over the decline in state revenues. Eventually they had to make cuts—drastic ones in many cases. To make up for the shortfalls, Californians passed new initiatives that afforded some protection of education funding, including the statewide lottery. But virtually all of these new funds went to K–12 education. Voters had little appetite for supporting the university and community college systems. As a result, state support of the university declined, forcing the universities to locate alternative sources of funding. One relatively untapped source of revenue was tuition and fees for student attendance. Abandoning its promises of a truly affordable college education, the university incrementally raised tuition, charging particularly large sums for out-of-state and international students, to the extent that tuitions began to rival the private universities and colleges. Yet the university lacked the sizable endowments enjoyed by elite private schools that could support students unable to afford tuition. In effect, at the same time that the university began to admit more students from less-privileged backgrounds, it started to charge more for attendance, not to mention shift budget priorities away from classroom learning.

Decrease in state support was closely entwined with changes in curricular emphases. Namely, education in the liberal arts was devalued relative to the STEM fields, decreed as "more practical." Supposedly some taxpayers resented their taxes going to literary criticism or study of the classics. At the same time, as students (and their parents) started feeling the economic precarity ushered in by neoliberal policy, they wanted to receive education in fields that would lead to relatively higher-paying jobs. Enrollment in the arts, humanities, and "soft" social sciences such as anthropology declined. Meanwhile, the university amped up its STEM offerings. UCSC, for example, did not even offer a degree in computer science until around 1980, and its School of Engineering, its first professional school, didn't open until 1997. By 2006 UCSC was offering a heretofore unheard-of degree in computer gaming.

Research institutions, like those in the UC system, had some strategic advantages over state and community colleges. More than seeking private donations from alumni and the like, they could obtain grants, contracts, and donor funds for research. The university began to push hard for faculty to fund their research with federal funds, to the extent that the ability to obtain grant funds became a consideration in reviews for faculty tenure and promotion, and sometimes a specific criterion. The university also encouraged direct partnerships with corporations, most infamously when UC Berkeley paired up with the Novartis Corporation to advance biotechnology research, representing one of the first programs in which a private corporation directly funded a university-based research program.[5]

Another new source of revenues, stemming from the university's research capacities, were patents on invention. Prior to congressional passage of the Bayh-Dole Act of 1980—note here the watershed year of neoliberalism—researchers were not allowed to patent innovations produced with federal funding. The rationale had been that publicly funded research should be broadly accessible to the public. Bayh-Dole's thinking was that not being able to collect patents squelched innovation. Thereafter, the university actively encouraged patenting through newly established offices of innovation, the purposes of which included ensuring the university got its rewards for housing the engineers who could now make a lot of money on the side. The obvious catch is that both corporate contracts and major sources of federal grants—such as the National Institutes of Health, the National Science Foundation, and the Departments of Education, Defense, and Energy—were rarely inclined to fund projects whose purposes solely advanced knowledge.

To the contrary, except in nominal amounts, most funding was awarded to projects that could promise eventual commercial utility. Owing to congressionally-mandated priorities, federal funders in particular sought projects that could enhance international competitiveness. Of course, only certain programs even carried the possibility of developing patentable inventions. As one manifestation of this new funding climate, the long held division of labor between basic and applied research eroded. Another is that faculty in non-STEM fields came to carry heavier teaching loads than those in STEM, since comparatively little federal funding existed to support their research. Meanwhile, to attract STEM researchers who could make more money elsewhere, the university began to pay higher salaries for engineers, along with allowing outside consulting and sharing patent revenues.

Again, the changes in the University of California system exemplify broader trends. As Elizabeth Popp Berman has argued, public universities especially had to reposition themselves as economic development engines to win back public support.[6] All but dropping their vestiges of science for science's sake and learning for learning's sake, universities promised instead that investments in the university would yield science that is economically valuable as well as produce the scientists who would innovate, create new industries, and hold high-paying jobs that contribute to taxes. That the university's primary role became producing outputs with market value, including scientists themselves, is the heart of what some call the "entrepreneurial university" and others call "academic capitalism."[7] A particular irony pointed out by *Science-Mart* author Philip Mirowski is that the more the university became embroiled in market activities, the more it lost its political justification for state support.[8]

Moreover, since essentially all public universities pursued the same strategies, they had to self-differentiate to draw money to their particular brand of enterprise. That is where branding comes in. For example, UCSC came to brand itself as "the original authority on questioning authority" even though the programming continues to move in an entrepreneurial direction. The key point is that in the university's quest to demonstrate relevance and the ability to solve problems in the real world, it began to take on the ideological and performative trappings of Silicon Valley. One important indicator was the embrace of "impact" as a key goal and metric.

## The Imperative of Impact (and Its Sidekick of Interdisciplinarity)

You can see the imperative of "impact" all over academia. It is in one sense an artifact of neoliberal audit culture, which uses all manner of metrics and performance measures to attempt to instill accountability and minimize waste.[9] Academic journals have "impact factors" that are literal measurements of how much that particular journal is cited, while university committees assess individual scholars' research output by the citations they amass. A major criterion for federal government research grants, on which reviewers of proposals must judge, is the "broader impacts" the research will have. Although these impacts can sometimes be framed around enhancing diversity, equity, and inclusion by training graduate students or other "human resources," which can be a good thing, mostly they revolve on developing physical, institutional, or informational infrastructure or, of course, developing and transferring technology. Students often feel the need to have impact too, without clarity of what kind of impact they're supposed to have.

In full disclosure I've been fortunate to obtain several grants from the National Science Foundation, and much of the research I discuss in this book is based on one NSF grant. I find the requirement to demonstrate broader impacts the most challenging part of the proposal. I'd like to think that the book you're reading now will change your mind and have an impact. So I might write in a grant application that I plan to write a book tailored for college students and young entrepreneurs about the importance of, say, doing research before you create a solution. But how can I possibly demonstrate that the book I write will actually change your mind? Will

I develop a survey of all my readers to see what they've learned? How could I possibly determine who they are in the first place? Much qualitative or critical social science simply does not lend itself to the kind of outcomes that can be quantified and reported. The fact that social scientists are funded at all by the NSF owes to the diligent work of program officers to defend programs in the social sciences that aren't necessarily instrumental, along with the peer reviewers who remind program officers that the material in these proposals can lend itself to broader social understanding.

In keeping with the objectives of the entrepreneurial university, the most prized impacts are those that can create direct economic benefit, making impact a proxy for return on investment. Particularly prized are solving "societally relevant problems" with technology and partners from other sectors.[10] "Impact" thus quite frequently takes the form of technological breakthroughs or science that will lead to them. This is quite different than connecting academic activity with issues and movements focused on ecology, equity, and social justice.[11] These latter activities take time and strategy, and immediate results are difficult to calibrate. They might deliver identifiable wins but not packaged solutions.

Defining impact as economically relevant, technological breakthroughs, or solutions further elevates the knowledge and applications of STEM researchers, while continuing to sideline the many artists, social scientists, and humanists in the university who also want to change the world but have very different ideas of how to go about that, perhaps even resting in response. This definition of "impact" particularly venerates engineering, an entire discipline that prides itself on solving practical problems. As it says on the UCSC engineering school's website, "We are problem solvers by nature, driven by curiosity and a desire to improve the modern world." Engineers are also most able to develop patentable inventions, even relative to other sciences that continue to conduct basic research. Again, social scientists struggle to demonstrate impact in these terms. As an executive dean from an Australian university put it: "The social sciences don't have the kind of technological breakthroughs that STEM can show that have transformed our lives. I think we [as social scientists] struggle to articulate the contribution that we make. . . . It's basically an existential state for the social sciences."[12]

That said, the university does have a way to manage the de facto marginalization of non-STEM faculty. It's called "interdisciplinarity"—another watchword of the entrepreneurial university and a major emphasis of funders as well. The conceit of "interdisciplinarity" is that individual disciplines cannot solve problems alone and that multiple perspectives are needed for any given problem.[13] Its emphasis owes at least as much to the appearance of giving everyone a seat at the proverbial table. In practice, most interdisciplinary projects give pride of place to the scientists, while the social sciences and humanists are invited along to confer legitimacy. Here I write as one of a collective of scholars who work at the intersections of science and technology studies and agriculture and food, and who have similarly participated as social scientists on STEM-based projects. Consistently we have found that our role in shaping technological development and related research is highly constrained. We are asked to assess willingness to adopt and public acceptance of new technologies, educate the public about the benefits of technology, and otherwise represent "society," but rarely are we welcomed in offering insights into histories of agro-innovations and their effects (e.g., the Green Revolution) or examining the underlying assumptions of technologies in-the-making.[14]

History has shown how critical it is to have social scientists and even humanists substantially involved in innovations *while they are being developed* rather than after the fact. That, in fact, is the rationale for the Responsible Innovation framework mentioned in the introduction. Consider the fate of genetic modification (GM) in agriculture, which remains a specter looming over innovation today. Social scientists have shown how public resistance to GM was less about the technology itself, as STEM researchers have imagined. It was more about its social and environmental consequences: privatization of seeds, corporate consolidation, and untested environmental consequences—issues that are often nonobvious to technical experts.[15]

The failure to account for societal values and needs early on in the design of an innovation can lead to unanticipated social and environmental problems of the sort that social scientists and humanists are best positioned to study, explain, and possibly avert if their expertise is taken seriously. Often, however, STEM researchers eschew what social scientists and humanists can offer and are even hesitant to engage discussions along the lines of Responsible Innovation. They tend to view social science input as slowing down and even impeding the forward progress of technology and thus thwarting the promise of solutions.[16] Some non-STEM researchers are willing to play along, because STEM research is where the funding is, and they become handmaids to projects that have already been conceived.[17] Many non-STEM researchers, though, are simply ignored. As a case in point, I return to the unfolding of the UCSC ag tech initiative.

## Case in Point

The introduction of the ag tech initiative at UCSC turned out to be an opportune moment to put into practice some of the principles alluded to in this chapter, to test the possibility for true interdisciplinary collaboration in which social scientists and humanists could substantively affect the direction of technological solutions before things were set in concrete. For that matter, to be able to bring to bear insights from my existing research on Silicon Valley agrifood tech seemed a vehicle for that project to show its "broader impacts" required by federal funding. In support of those aims and more, I brought on board a graduate student to conduct interviews with various campus constituencies about the direction of the initiative. Hoping that cross-campus engagement could affect the direction of the initiative away from potentially adverse consequences and hence better solutions, I suggested the campus hold a symposium to discuss different perspectives on ag tech. Initially this idea met support from key administrators. Unfortunately, the results of both activities proved the case.

## The Interviews

The interviews revealed significant skepticism that such an initiative could successfully build on and combine distinctive campus strengths in agroecology, engineering, critical social science, and an overarching commitment to social justice. We found great disciplinary differences in how faculty approach their research and address problems, collaborate in interdisciplinary teams, and incorporate social justice into their work. We also found significant divergence in appreciation of the utility of ag tech for agricultural problem-solving, ideas of who should be involved in the initiative and who it should benefit, and whether ag tech could meet the needs of small-scale farmers. The few points of agreement were actually sources of tension. Many respondents felt the initiative to be a forced marriage and that uneven funding across divisions was a barrier to collaboration.[18]

With details found in our ensuing publications, here I want to draw out three differences that illustrate how the university's empowerment of engineers relative to others could easily lead to the kind of solution-making in agriculture and food this book has problematized. It all begins with how the different disciplines define and approach problems. In these interviews both social scientists and agroecologists saw food system problems as deeply entwined with the social world and questioned the extent to which technology could address these problems. To quote one agroecologist:

> A lot of the persistent problems that we have within the agricultural and food system are things that can't be easily solved. Thinking about issues of food access, hunger and distribution, and equal access of wealth.... Those are the real persistent and wicked problems, and none of those are going to be solved by a technological fix.

Engineers, conversely, reiterated their discipline's singular approach in which they identify problems in relation to technological solutions they could provide. One said: "As an engineer, if there's a problem somebody has, and I have a technology that can provide a solution, that's what I do." Or as another put it: "If you aren't solving problems, you're just in the frickin' mud."

The engineers didn't only favor addressing those problems amenable to what technology could fix, to the exclusion of social concerns. Even in relation to farm-level production issues, they approached the inherent biophysical complexity in a more narrow way, focused not on exploring the problems but orienting around a predetermined goal. Engineers emphasized the importance of zooming in on specific phenomena and processes that could be addressed, a kind of problem closure. One specifically noted the largely unpredictable nature of complex biological systems as their biggest challenge, impeding their overarching goal of increasing system efficiency. For agroecologists, that complexity is the point. Describing how they undertake research, one agroecologist said: "You need to understand the community that you're embedded in, and the impact that your agriculture has either on or within that community . . . from production to our use of resources, potential pollution, things like that . . . it is a very whole-systems perspective." Agroecologists emphasized the complexity of biological organisms and the use of practices that further foster species diversity and "interactions between plants, soil, microbes."

Third, the interviews shed light on why engineers found little use for social science expertise. Reflecting a disciplinary propensity for internal reflection, a social scientist voiced awareness that their input is often unwelcome and dismissed as overly negative, but rather than "being cynical and skeptical about everything," they were guided "by the radical optimism that there are other ways of doing things." In contrast, most of the engineers had little to say about what social scientists could offer at all. One did note that social scientists are interested in the impacts of technology on farmers but admitted to being just "really concerned about efficiency."

Based on these snippets, you can certainly get a sense of why the neoliberal university favors engineering. Engineers have a can-do sensibility that looks for problems they can solve, with the tools they have at their disposal. In contrast, social scientists and even agroecologists tend to make problems ever more complex in ways that may seem paralyzing. By the same token, engineers seem to see almost all problems as related to inefficiency or at least the ones they can solve. As I have suggested throughout this book, efforts to bring more efficiency to agriculture are precisely what led to technologies that ultimately damaged environments and never really helped farmers. The "radical optimism" expressed by that one social scientist suggests that problems should be explored, not predetermined, precisely so "other ways of doing things" can emerge and be nurtured.

In short, these interviews illustrated two very different mentalities for addressing problems in agriculture. While engineers embraced solution-making as core to who they are, the others suggested something more like response, beginning with the question of what a system needs. Unfortunately, the symposium only accentuated these differences and, worse, suggested little possibility that the concerns of agroecologists and social scientists would be heeded in further development of the ag tech initiative.

## *The Symposium*

Planning for the symposium, which took place among a small committee, revealed additional tensions. Repeating at several junctures that they just want to solve problems, the engineers bristled at the proposal to discuss epistemological or methodological differences among the different disciplines as a way to gain better understanding of what each group could bring. Indeed certain committee members squashed all suggestions that would involve drawing out critique or differences in favor of a format that assumed that the initiative would move forward. Interested faculty and graduate students would give four-minute lightning talks on their ag tech areas of interest (strikingly reminiscent of a Silicon Valley pitch event). These talks would be followed by breakout groups focused on areas that attendees would be interested in pursuing, areas of strength in ag-tech that are worth supporting or building upon, and valuable next steps in developing more ag tech work at UCSC.

Despite this forward-looking, dissent-squelching, and practical approach, the event further revealed that campus constituents were not on the same page. The lightning talks highlighted fairly disparate understandings of what ag tech could be at UCSC, with some presenters showcasing their work on autonomous vehicles, others discussing the utility of a bee hotel, some promoting their particular facility or program, and still others hinting at the potential for unintended consequences along the lines of the Green Revolution. The breakout discussions that followed were even more contentious, precipitated by a presentation of our report's findings, as well as a robust coordinated presentation by four agroecologists. Recounting a "cautionary tale" of UC's highly contested development of a mechanical tomato harvester in the 1970s, they shared a quote from a lawsuit that followed: "Not a penny was spent to study the implications of mechanization on jobs, on farm sizes, on prices, on the environment."

During the breakout sessions, some attendees expressed excitement about the initiative and followed the session prompts; many strayed from them and expressed that the symposium's proceedings exemplified the disjuncture between the administration's aspirations and faculty expertise and interests. Many attendees thought that sustained, iterative, and respectful conversation was most needed to achieve meaningful buy-in and collaboration. Since then, and to this writing, no such discussion has taken place. The administration continued to tout ag tech as a huge opportunity, and engineers appeared to move forward with their plans, undaunted by and uninterested in others' disciplinary knowledge.

## Sidelining the Social

Perfectly exemplifying the imperatives of the neoliberal university, the ag tech initiative promised new revenues in grants, contracts, and patent royalties and the ability to show relevance and impact, albeit narrowly defined. The gamble was that the university would maintain its historic reputation in social justice and agroecology—that is, its branding, without the substantive input of the very scholars and programs that have built that reputation. And that is what "impact" looks like in the neoliberal university, now a progenitor of Silicon Valley–style solutions.

Social scientists who might bring depth to problem analysis, who might help avoid the mistakes of the past (and present!), *who might avert the problem with solutions* can't even get in the door, while those who want to move swiftly with the brightest and shiniest technologies, regardless of their social assumptions or implications, are given the helm. In other words the pursuit of solutions at the university reinforces techno-fixing while leaving little room for response. Is there any wonder why agrifood solutions continue to be misguided? What is strange in this context is that more than Silicon Valley, which has no institutional need for interdisciplinarity and the expertise of humanists and social scientists, the university has the capacity to avoid some of the problems with solutions. It has at its disposal scholars who know histories of technology, who analyze the socioeconomic dynamics of food and agriculture, who study how transformations happen in this domain.

And yet, as I discuss in the next chapter, not only is such expertise sidelined; Silicon Valley culture has so infused the university that it is now in the business of producing solution-makers who act with the same hubris as the techies, with little opportunity to learn from the past—or even their own mistakes.

# 7 Big Ideas and Making Silicon Valley-Style Solution-Makers

*We've got some problems—problems that need new eyes. We know, you're busy: class, term papers, research, work. There's a lot going on. But you're not just a student. You're a visionary, an entrepreneur, a designer, an innovator, a connector, a builder. We know you've got what it takes: energy, truth, insight, imagination, and ideas. Listen, if you can dream it, you can create, launch, fund, and scale it. At Big Ideas over 6000 students have launched over 450 innovations, in 50 countries. So you bring the idea, and we'll bring you the structure, mentorship, networks, funding, and validation to get you going. No idea is too small; let's figure this out together.*

These are the lines, delivered with pregnant pauses, of a two-minute video (once) featured on the website of the Big Ideas contest, held at the University of California–Berkeley every year. Founded in 2006, Big Ideas, according to its current website, is "an innovation ecosystem that provides training, networks, recognition and funding to interdisciplinary teams of UC Berkeley students who have transformative solutions to real-world problems."[1]

The Big Ideas contest at Berkeley is one of hundreds, if not thousands, of university-based or university-oriented programs that encourage students to create solutions to some of the world's most challenging problems. Programs offer tracks in solution areas such as global health, environmental sustainability, antipoverty, energy, education and literacy, housing, and of course food and agriculture. Indeed there are an increasing number of programs training students to be solution-makers in food and agriculture alone. The marked increase in university programs oriented toward solutions correlates closely to and sometimes overlaps with the proliferation of programs in civic engagement and service learning designed to have students step outside their role as learners, to become doers instead. Standing apart from programs that have practicum attached to academic studies, these university programs instead seek to empower students to make changes outside of the confines of classroom learning.

For their part, students are eager to take part in such programs. They know that things are really messed up and they want to make an impact but are not sure how. Such programs empower students to see their ideas as worthy but at the same time provide a structured opportunity to take concrete steps toward "real world" change. And participating in these programs doesn't exactly hurt for getting a "real job" once students graduate. The problem is that many such programs, although by no means all, are visibly modeled after Silicon Valley style. These programs have inherited the mantra of "doing well by doing good" and are convincing young people that they will have more success if they use the tools of capitalism rather than public action to make needed change.[2]

To give a sense of their proliferation, Enactus.org, which provides a platform for entrepreneurial programs for "sustainable change," claims more than seventeen hundred participating universities worldwide; a quick internet search suggests such programs abound far beyond those numbers. More to the point, many such programs are structured to replicate the ecosystem of incubators and accelerators that bring the "doable" solutions. Borrowing most closely from the Silicon Valley model, they culminate in pitch competitions. Indeed, on some campuses these programs are nothing more than pitch competitions. In direct imitation of Silicon Valley style, university pitch events ask students to design a solution to a major challenge and submit that to judging criteria emphasizing the potential impact of the product or program, along with its viability and creativity. Rather than venture capital funding per se, contest winners generally receive prize money and other forms of support to bring their ideas to fruition. This helps them, per Big Ideas, to "make an impact all over the world."

Due to this structure, student ideas and approaches are unlikely to fall too far from the Silicon Valley tree. To be sure, a quick review of some of the winners on the Big Ideas website shows solutions with certain technical novelty (especially from the STEM students) but that are otherwise well within the menu already on offer. For example, winners in the agriculture and food category include digital farming, alternative protein, nutrition platforms, hydroponic farming systems, and plastic recycling, to name a few. Even the occasional solutions more aligned with nonprofit, nontechnical models—for instance, recycling campus food waste to support food insecure students—are not all that novel. Still, my concern is much less about the novelty of the solutions. While "novelty" is often a criterion of judging for these contests and many other innovation prizes, the search for novelty can actually detract from what most needs to be done. Rather, it is that, as you might infer from the language of the Big Ideas video quoted at the outset of the chapter, they instill in students the Silicon Valley approach to solution-making, hubris included. And they do so without even bringing to bear the benefits of a liberal university education.

## Pedagogy of the University Pitch Competition

In 2016, out of peculiar curiosity, I agreed to serve as an expert judge for the Big Ideas contest at Berkeley. That particular contest takes place in three stages. First, contestants submit a three-page pre-proposal. Those who make the cut to become semifinalists submit longer written proposals and a short video. The few who become finalists participate in a live pitch event. I was assigned to review ten or so pre-proposals in the "food systems" category and rate them among several dimensions. I found the ideas less than big; several proposals, for example, featured some sort of hydroponic system or vertical farming structure for at-home or community use. Still, what struck me more was the near absence of a problem analysis relative to the technical specifications of the inventions. At best, the proposals stated a factoid or statistic about, say, land shortages or food insecurity, but otherwise lacked evidence that students had been prompted to explore a problem in any depth as a precursor to developing an appropriate solution. To the contrary, it appeared that the emphasis was on specifying the "big idea." And so the problems appeared in typical solutionist fashion, as after-the-fact justifications for a solution already conceived.

In addition, since virtually all the proposals forwarded a technological solution, the after-the-fact justifications were typical of the techno-fix—a problem that could be addressed with technology rather than politics. Many proposals assumed that the problem lay with some sort of shortage rather than, say, an absence or loss of entitlements à la Amartya Sen. Of course, much like Silicon Valley entrepreneurs, these students were likely not in a position to execute a political solution even if a modest techno-fix was clearly not equal to the task. But that doesn't make the solutions appropriate. For that matter, none of these proposals discussed how the solutions would be introduced and disseminated in the communities targeted for the solutions, much less whether these communities would be consulted in advance. In effect, the parameters of the contest encouraged students to contemplate arguably neocolonial ventures, reflective of the will to improve, as if the students would know better about how to manage food production than even experts in the community. In other words, the proposals were rife with the problems with solutions.

Based on this initial experience, I became interested in the pedagogy of these pitch competitions, both in terms of the substantive learning they might offer or require as well as what values and approaches they attempt to instill in participating students. How do these programs imagine that students learn about the problem, who it effects and how, and how similar solutions have been implemented and been received in the past? What training do students receive in the eventual dissemination of their big ideas, whether the ideas take the form of technologies to introduce to "users"—a very techy term—or to their targeted communities? What curriculum in other words, must students follow before they embark on saving the world? After all, the problems with solutions I saw in those pre-proposals were not preordained, but enabled by those who designed the contest. As university-supported endeavors, these programs ought to take seriously the learning experience.

Based on what can be gleaned from websites, the preparatory programs of some of the most high-profile competitions are quite limited. MIT, for example, advertises its annual Food and Agribusiness Innovation Prize as the premier business plan and pitch competition for US university and graduate students developing new products or technologies to improve sustainability in the food system. Although students likely come with strong technical backgrounds, the program is open to all, without prerequisites. The program seems to entail pairing students with mentors in the development of a business plan. Likewise, the Friedman Food & Nutrition Innovation Prize offered at Tufts University (a center of food and nutrition studies) makes little mention of preparation other than attendance at educational workshops and the opportunity to engage in one-on-one coaching and pitch practice.

The Tufts judging rubric, meanwhile, looks like it was lifted out of a Silicon Valley pitch competition. Questions include: "Does the team understand the problem and target customer? Does the proposed solution solve the problem in a unique manner compared to the alternative? Does the team have a credible and thoughtful go-to-market strategy? Is the proposed business model financially sustainable?" "How is the world a better place if the team solves this problem? Did the team convince the judges they are the right people to solve this problem?" As with MIT's contest, there is no emphasis on curricular requirements before participating. The presumption seems to be that participating students will receive requisite technical training through their course work. Through that lens the Big Ideas contest seems to be one of the more evolved of such competitions.

*Big Ideas*

With that in mind, let's take a closer look at the Big Ideas contest. Drawing on material available on its website in 2022, the contest is by far the most elaborate in terms of program rationales, application processes, and preparatory work. The first thing to note is that the contest is not funded by the university but by private funders. This includes private foundations and the Blum Center on Developing Economies, endowed by the late investor and UC regent Richard Blum. These funders began the program to support students, according to the website, "in making social change—whether through for-profit endeavors, nonprofits, or small ad-hoc teams." To do so, the contest asks students "*to step outside of their traditional university-based academic work*, take a risk, and use their education, passion, and skills to work on problems important to them."[3] Here you can already see the trappings of the neoliberal university: a donor-driven effort aimed at practical learning and immediate problem-solving, at potential odds to learning through critical analysis. In fact, such philanthropy-driven education programming is part of a much larger trend, often operating with little accountability but having undue influence on curricula.[4]

At the same time, the Big Ideas contest truly does emphasize transformative change. Moreover, it has significantly evolved since I served as judge. According to its website, the Blum Center has continually modified the design and management of the contest, based on "continuous reflection and iterative change." So satisfied with its development of a "proven, replicable model for managing innovation contests on university campuses," the center offers a toolkit of "proven contest management strategies," along with "lessons learned, best practices, and honest reflections on the process of managing a student-led innovation contest." All of these aspects make Big Ideas an optimal vehicle to assess the pedagogy of a competitive, entrepreneurial approach to world-saving. The program requires that projects are early-stage ventures and have a clear social impact, "centered around an innovation (technologies, services, programs) that produces a clear benefit with the potential for sustained improvement in the lives of groups or individuals." However, the program also requires that projects be initiated and led by currently matriculated undergraduate or graduate students and "cannot be an extension of faculty-guided research or *led by a nonstudent established organization*."[5] That means that students do not have the opportunity to learn from others already doing the work.

In terms of actual programming, the Big Ideas contest website acknowledges the complexity inhering in its key impact areas and challenges applicants to reflect on them. For instance, the description of a workforce development track asks students to think critically about how new technologies in AI and robotics could displace workers and suggests that solutions "help these workers regain not only their lost income but also their purpose and direction in life." The global health track asks that proposals "demonstrate evidence of a widespread health concern faced by resource-constrained populations, and develop a system, program, or technology that is culturally appropriate within the target communities and designed for low-resource settings." The description of challenges in the food and agriculture track is a little less evolved, in my opinion. Yes, it calls for innovations and solutions "that address complex challenges in food systems and agricultural development," but its areas of focus include "enhancing agricultural production, increasing food security, promoting sustainable farming practices, and/or creating equitable access to nutritious food." I hope you've gathered that if "enhancing agricultural production" means increasing yield, that can work at odds to those other, more laudable goals. (If not, please return to chapter 2 for a review.)

The content suggestions for the three-page pre-proposals also reflect an evolution from when I served as judge in 2016. As of 2022, the application asks for at least half a page on a "description of the problem or need that the project will address, communicating an understanding of relevant research/statistics on the problem." At least this is a start in encouraging students to see the problem statement as more than a throwaway line. It also asks for at least half a page on "a landscape analysis consisting of an overview of any services, programs, or products that have already been designed or implemented to address this problem and addressing the strengths and limitations of these approaches, as well as the gaps that still exist." The application suggests the remainder be devoted to summarizing the innovation (e.g., project, service, or product)—including "how it differs, how it works, and its potential for social impact." The requirements allot additional space for team biographies and references cited.

Finally, relative to other contests, Big Ideas appears to give more effort to student training and mentoring "to turn their ideas into action." The website mentions a yearlong process in which students develop skills in critical thinking, market analysis, team-building, and presentation. So far, pretty good. But if you look at the training program more closely, you see its narrowness. In practice, the training consists of a few workshops and panels, offered to the entire field of applicants, on problem framing, design thinking, and landscape analysis. Students who survive the very competitive first round cut go on to attend writing and story-telling workshops, and they also have a session with a mentor experienced in the field. As such, the content of this training draws right from the Silicon Valley playbook: by the very definition of these terms, students learn how to *perform* pitches rather than learn substantive content.

For example, "problem framing" connotes presenting a problem to be compelling, not learning to research a problem to understand it. Students might learn that their pitch will be strengthened if they show some knowledge about the field in which they enter. But if that means throwing in a few statistics here and there about a massive problem, as routinely takes place in Silicon Valley pitches, student knowledge can remain thin indeed.

Likewise, design thinking recommends steps to perform the solicitation of potential user input but does not dictate serious engagements with communities who might be affected by a new design. Story-telling chops are the standard fare in books on pitching; entrepreneurs with good stories and compelling presentations are often awarded with venture capital funding. (Recall Lisa Curtis's legendary story of encountering *kuli kuli* as a Peace Corps volunteer in Niger, recounted in the preface.) So the skills students gain are in *performing* impact but not necessarily *making* it—skills that may attract funding and support but do not exactly change the world.

To that concern, what I find missing is any mention about aligning students' big ideas with knowledge they might gain through academic work. Big Ideas, along with other contests, has no requirements that students complete coursework in the social, cultural, and political economic contexts of their proposed work nor, for that matter, that require students to reflect on their own social positions as changemakers. To the contrary, students are asked to step *outside* their traditional academic work, not to bring it to bear in their big ideas. In effect, students are encouraged to act, without acquired expertise. When "students aren't experts in anything," as one critic of the use of design thinking in a college curriculum put it, the projects often take the form of "kids trying to save the world."[6]

## The Effects of Big Ideas

Perhaps you think I am making too much of pitch competition curricula, or lack thereof. They are empowering to students and bring emotional satisfaction. So who cares about the intellectual work that may or may not go with these programs? Besides the fact that they are taking place in institutions of higher learning, which should mean something, you have to consider the stakes of impact competitions especially relative to classic venture competitions. It is one thing to invite students to develop and compete over business ideas for products or services that might not find a market; it is quite another to invite them to develop and implement interventions for improving other people's lives. If they do not learn about the origin and character of a problem, its situatedness in time and place, how can their solutions be appropriate to the problem? If they haven't reflected on their own positionality and the neocolonial impulses of fixing the world, how can they expect to have their solutions welcomed and desired? While unintended consequences are always hard to avoid, a curriculum that deliberately omits such considerations gives students a jump start on them. To recall the words of Teju Cole on the White Savior Complex, "if we are going to interfere in the lives of others, a little due diligence is a minimum requirement."[7]

I suppose the good news is that in a competition format only the winners get the opportunity to take their interventions on the road, thus containing the possibilities for widespread undesirable consequences. Yet the effects of a Big Ideas-type pedagogy are much more diffuse than on the communities in which the winners will eventually take their solutions. It goes without saying that most direct beneficiaries are the student-participants themselves;

participation in Big Ideas and all manner of student engagement programs builds résumés and provides connections for future student success. The self-servingness of these programs is not the whole of it, though. Rather, these entrepreneurial, tech-oriented change programs teach lessons to students who genuinely want to do good, the many students out there, who for understandable reasons are dissatisfied with, and even outraged by, the social and ecological status quo. Yet the lessons these programs teach are not responsive to the ecological and social challenges students face.[8]

For one, these contests teach that social change is a competitive endeavor. By definition, contests have students compete over ideas rather than collaborate to generate the best ideas. The very notion that competition generates the best outcome is another sign of the infiltration of neoliberal logics into the academy—in an institution that should, of all places, value collaboration and learning from others. Second, these contests encourage students to generate a novel idea without benefit of working with those on the ground who may already be endeavoring to effect change. Not only might students be dredging up ideas already dismissed, or just not suitable, they are not getting the opportunity to learn from organizations and institutions who know their field of change and, optimally, work closely with the communities to which they will bring their solutions.

Third, the contests reinforce an ethic of paternalism rather than solidarity, versions of Tania Li's "will to improve" or Teju Cole's White Savior Complex.[9] At first glance, seeking to empower others seems right and generous. Yet by inhabiting the role of the one bringing empowerment, students necessarily do so based on their own ideas, even "big" ones, of how others should live and believe their expertise to be superior to those they wish to benefit.[10]

To the extent that such programs cater to students' needs for affirmation, as the Big Ideas introductory video most certainly does, these programs reinforce students' entitlement to act on behalf of others.[11] Again, participating students are unlikely to have the expertise of the third world trustees that concern Li, but by virtue of participation in these programs, students can get the impression that their ideas are superior to those who live in the conditions in apparent need of change.

Notably, even programs encouraging (or allowing) nontechnological solutions can foster these sensibilities. Not always having a grasp on what else they might offer, some students gravitate toward educational programming in hopes that teaching those with less privilege will elicit changes in behavior. The unacknowledged assumption is that poor circumstances are a result of their personal choices rather than social structures.[12] This is an impulse I have seen with my own students who, until disabused, want to teach people how to grow, cook, or eat food rather than organize with them to change the conditions of food production and consumption. In contrast, programs emphasizing technological solutions engender the sense that you can't change people's ways and you therefore need to change the technologies around them—the classic techno-fix. Believing in the value neutrality of science and technology, techno-fixers may even be unaware that they effectively build their own values and assumptions into technologies.[13] Still, like those drawn to educational programming, techno-fixers learn to take for granted that their chosen improvements are universally desired and unquestionably good.

But again, the fault of the will to improve or the White Savior Complex is not only about the unintended reproduction of harm for the intended beneficiaries, but what it does for the white saviors. As Cole tweets, "the White Savior Industrial Complex is not about justice. It is about having a big emotional experience that validates privilege." What privileges do saviors enjoy? The privileges of not relinquishing their own privileges, of not recognizing their own complicity in systems of oppression, of setting the terms of what others will get or need—and, yes, the emotional satisfaction.[14]

This brings me to my final point and biggest reservation with the lessons of these competitions: they convince participating students that they are doing something important when they are doing things other than the difficult change work that needs doing. Competitions like Big Ideas, in short, encourage innovations that attempt to put a lid on a specific problem rather than draw on collectivities to reimagine what the world could look like and how to get there. These programs give students false hopes that their solutions will be impactful without giving them space to come to terms with the world they face.[15] They do not create activists who have the capacity and willingness to organize for the long haul, and instead arguably suck the air out of more activist impulses. Indeed, these competitions are depoliticizing for issues that desperately require political responses and promise fixes that do not disrupt the systems that created the problems in the first place.[16]

Perhaps none of this is all that surprising for a competition funded by investors who want solutions and not fundamental change, and who like to produce others in their image. Yet to fault the program solely for its market orientation misses something fundamental about a pedagogy that encourages just doing. Many of these problems could be tempered, if not entirely avoided, if the pedagogy made space for knowledge—if the pedagogy, that is, weren't so anti-intellectual as well as apolitical. Moreover, these programs might cultivate better changemakers if they made space for analysis. Instead, the curricula seem to end at the moment of the pitch, giving students no opportunity to reflect on the effects of their solutions, even if hypothetical, to consider what they could do differently in the future. These pitch curricula in that way miss a key element of "praxis"—a form of critical thinking that comprises the combination of action *and* reflection.

### The Case for a Praxis Pedagogy

The notion of praxis is most associated with the writings of Paolo Freire, a Brazilian educator and philosopher who was a leading advocate of critical pedagogy. Freire called for action *and* reflection, "directed at the structures to be transformed."[17] Champions of praxis see it is an ongoing, iterative process of learning and doing, not a separation between the two. Critical to praxis is attentiveness to how problems are constructed and by whom. Those engaged in praxis never assume universal agreement about the existence of problems much less the character of them. Equally important is critical reflection on past actions, including consideration of who and what was served by these actions so that change agents might do it better next time around. Praxis also sees doubt and uncertainty as chances for learning, not emotions to suppress in the name of performing confidence. Finally, praxis implies an ongoing commitment to social change, not implementing a solution and moving on.[18]

The pedagogy of Big Ideas and other competitions I reviewed couldn't be more different than a pedagogy based in praxis. Big Ideas–style programs emphasize the singular problem, the right solution, and the plan—and judge on those criteria. Ironically, the

Blum Center has apparently taken the time to reflect on how it could do better, as indicated by its language about the toolkit, but it does not appear to offer the space for students to undergo the same reflection on goals, means, reception, or any other way to systematically learn from the experience. A solution that pitches well may not play out well, and although mistakes will be made, repeated mistakes can be avoided. In my view we should be skeptical of a kind of "learning by doing" that does not account for its consequences.

The thing is there *are* pedagogical models that involve students in the world of social change utilizing the principles of praxis. Models based on praxis foster curiosity about the context and character of problems, teach humility rather than hubris in social change efforts, encourage solidarity rather than neocolonial correction, and ideally make young people lifelong activists of some kind or another. Such approaches embrace rather than sideline critique as an important part of social change—and learning too. I have taught in a program based on a praxis pedagogy for more than twenty years. It is the Community Studies major at UC Santa Cruz, and its curriculum serves as a true antidote to the Big Ideas pedagogy.

Founded in 1969, the Community Studies major was designed for students who wanted to do academic work but also be involved in world-changing activism. From its inception Community Studies has centered the integration of academic classroom work and fieldwork, a characteristic that differentiates it from the many programs that involve internships or experiential learning detached from an academic component. The curriculum itself revolves around a six-month, full-time field study during which students work for social justice or social change organizations. Some of the first field studies involved working with the United Farm Workers' boycotts; these days students are more likely to work in health clinics, community gardens, food policy, LGBQT+ support networks, or housing advocacy organizations. No matter what, students must work with established organizations that are already working in areas of student interest. Unlike the Big Ideas approach, the Community Studies program is committed to having students learn from and about organizations that have experience on the ground.

Before students go to the field, they have to undertake substantial preparatory course work—that is, they can't go solve things without prior knowledge. Although the Community Studies curriculum has changed over the years, today it involves no fewer than three courses in substantive, topical areas related to the student's future field study. The idea is that students enter the field with some understanding of the nature of the problems they will be working on, the communities in which they will immerse themselves, and the kinds of social action that have occurred in the field. In addition, students take a course each in community organizing and the nonprofit sector, so they understand the institutional terrain in which they will be working, including the limits of the "nonprofit industrial complex." This refers to the funding model for nonprofits that ultimately is based on the wealth of benefactors.[19] Crucially, students also take a course that prepares them methodologically for field study. This includes lessons on taking field notes, making themselves useful, and deep reflection on their own positionality and ethical commitments as they enter the field.

The field study itself is completely immersive, involving study of and participation in a community engaged in building social justice and working for social change. It is this—not the pitch—which is the central feature. Students take daily field notes and complete papers on their experience for the course credit they receive while in the field. Finally, when they return from the field, students take a capstone course that requires them to come to grips intellectually with that experience. In analyzing what they learned, students generally consider the obstacles and possibilities for social change in their specific arena, recognize the place of politics, and rarely imagine a quick solution. The analytical work students do during field study and following is critical to the process; the experience ends with honest reflection on "the politics of the possible."

You might find it paradoxical that the Blum Center at UC Berkeley, which supports the Big Ideas contest, also helps support a program partially modeled after Community Studies: the Global Poverty and Practice minor. Focused solely on poverty and inequality, this program asks students to think critically and immerse in anthropological and historical scholarship. After such preparation students engage in a six-week practicum, in which they connect theory with action by partnering with nongovernmental or community organizations, government agencies, or other development programs domestically or abroad. In a book reflecting on the experience of teaching in the Global Poverty and Practice minor, founder Ananya Roy and others write that "students in the program were not meant to be vanguards of social change: they were not billed as global leaders; they were not given the charge to solve urgent global problems. Instead, they were invited to work modestly, reflexively, and persistently with both marginalized communities and communities of inquiry."[20] Indeed, the program teaches that action-based orientation "is part of the problem," as "enthusiasm can be dangerous and good intentions can be deadly."[21]

Both the Community Studies major and the Global Poverty and Practice minor share a conviction that critique has an important role to play, that finding the perfect program in which to work misses the point, that trying to eliminate contradiction is fruitless, and that it is okay and even productive to express doubt about your approach. Both programs encourage students to recognize their complicity in systems of oppression and to walk the line between "the hubris of benevolence and the paralysis of cynicism."[22] Finally, both programs refuse the anti-intellectualism encouraged by programs without an academic component and insist on reflection as a mode of action—in other words, doing by learning. These two programs, I want to suggest, value different sensibilities than the Big Ideas contests and other such competitions that take their cues from Silicon Valley. To the contrary, they cultivate the very qualities that a group of higher-education scholars have termed as "education otherwise":

> Education otherwise offers no predetermined, prescriptive answers, because it operates as a life-long inquiry into the challenges and complexities of learning/unlearning, including learning from failure. Instead of encouraging investments in an idealized future, education otherwise encourages us to situate hope in the quality and integrity of an ongoing practice of expanding our capacity to face and process difficult intellectual, affective, and relational challenges in the imperfect present. It also asks us to take into account the past and how it has shaped the social and ecological injustices that have accumulated and become sedimented within the present.[23]

In short, such programs provide tools for response not solutions, and I would argue they cultivate the kind of changemakers necessary for the current moment. I expand on this point in the conclusion.

# Conclusion

## The Pessimism of Solutions and the (Cautious) Optimism of Response

One of the more visible figures in the tech sector is Eric Teller, who goes by the name of "Astro" Teller. Also known as the Captain of Moonshots, Teller works for the company X, formerly Google X (and not to be confused with the former Twitter), a quasi-secretive subsidiary of Google's parent company, Alphabet. The putative mission of company X is to pursue technologies that are so radical, they can solve humanity's biggest problems. Teller likens these technologies to moonshots, and in his many appearances at conferences and in print media, he has made the point that a true moonshot is laughable. If an approach sounds reasonable, he often says, they are simply not moonshots.[1] Given X's track record with technologies such as self-driving cars, balloons that provide internet in rural areas, delivery drones, and a technology to store electricity using molten salt, while abandoning their earlier attempts to create carbon-neutral fuel from seawater, you might wonder how X defines humanity's biggest problems.[2] Nevertheless, that X even exists illustrates Silicon Valley's continued investment in techno-optimism.

## A Tale of Two "Moonshot" Solutions

The idea of moonshots obviously draws its inspiration from the first successful US moon landing in July of 1969 and is frequently deployed in the tech sector to galvanize investment in technologies that may not prove feasible, yet won't stand a chance of coming to fruition at all without an influx of cash. Air Protein is one such technology, and it is not entirely coincidental that it is based in technologies originally developed by NASA for space travel. Yet to be commercially released, Air Protein is a solution to conventional livestock production, impending protein shortages, and the climate crisis all at once. What makes Air Protein laughable, to borrow Teller's metric, is that it is "meat" made with a genuinely far-fetched technology. Air Protein claims to involve fermentation of microbial substances derived from air to have a "carbon negative" effect.

Then there is Moonshot Snacks, a brand of Patagonia. Moonshot Snacks are an offshoot of a carbon management software platform called Planet FWD, yet another technology that provides information about carbon footprints and how to reduce them. But Moonshot Snacks are far from high tech. They are crackers made with ingredients from farms practicing regenerative agriculture. "Regenerative agriculture" is a new trend in sustainable farming, not all that distinct from organic, that involves soil health practices that presumably remove carbon from the atmosphere. Moonshot Snacks are packaged with recyclable materials, contributing to their "net positive outcome." With nothing obviously outlandish about these snacks, the founder claims to have opted for branding these snacks as moonshots to suggest "uplift."[3]

The aspiration of Air Protein is laudable. Its developers do aim for the sky, as it were. Indeed, the point that we should unleash human creativity in service of radical problem-solving is highly compelling. What makes Air Protein laughable in my view is the presumption that an über techno-fix is the optimal solution relative to many other possible, and far less technically challenging, responses to the problems it is supposed to solve. As for Moonshot Snacks, the practices that the brand supports, both recycled packaging and agricultural techniques founded in agroecological principles, are good ones. What makes Moonshot Snacks laughable is the premise that selling crackers to conscientious consumers will change the world. Rather than looking to either the über techno-fix or consumer snacking decisions, imagine instead if we drew on the aspirations of a moonshot to create the social conditions in which these agroecological principles were widely put into practice!

Brought together by the coincidence of the moonshot moniker and undoubtedly both guided by the will to improve, these two products set the range of what the tech sector is offering in terms of solutions, here specifically in the realm of food and agriculture. On one extreme are extraordinarily aspirational techno-fixes supported by misconstrued problems. On the other end are the ultimate in low-hanging fruit, repackaged renditions of already existing alternatives to industrial agriculture in edible form. Given either moonshots' lack of attention to the political and economic conditions that have given rise to malnutrition and climate catastrophe, neither product can rise to the task of confronting a deeply unequal, unsustainable, and messed-up world. Nor can many of the "solutions" between these two extremes and those in the other domains where solutions big and small are being conceived. And yet these solutions sideline other modes of changemaking and practice that have a shot at addressing the serious problems of food and beyond.

## The Pessimism of Optimism

In June 2023, as I was finishing this book, I came across a multipage pull-out advertisement for the Collab Fund in the *New York Times*. In its most prominent text, the ad read:

> There is no bigger problem, no greater challenge, and no greater opportunity than addressing climate change. Throughout history crises have two things in common: Solutions only come when the cost of inaction becomes painfully obvious; and the amount of ingenuity solutions bring exceeds what anyone imagined. We are optimistic about our shared future.

There it was: an artifact of the problem with solutions that I just happened upon during my Sunday morning routine, arriving in the form of an ad for a venture capital fund that "backs companies that live at the intersection of for-profit and for-good." Look at the celebration of human creativity. See the invocation of optimism. Imagine the will to improve and the emotional satisfaction of doing *something*.

So what exactly is the Collab Fund optimistic about? Based on the selection of companies featured in the ad, a variety of techno-fixes, some obviously based in a solutionist approach: AI and robots to identify, sort, recover, and reuse more raw materials from waste streams; biodegradable and compostable alternatives to plastic; smart shipping containers that don't use chemicals or refrigerants; filters of excess $CO_2$ that measure what is captured and made into "high-integrity" carbon credits; portable toilets that turn bodily waste into useful fertilizer; data analytics of user needs to "empower" utilities to avoid electrical grid stress; design and installation of residential thermal systems; and, of course, a host of food solutions, including "tasty and organic plant-based snacks." Some of these techno-fixes are more compelling and novel than others, but they are all products. Products that can be sold for profit, which is why they appear in the portfolio of the Collab Fund. But is it only that feature that renders them problematic solutions?

It would be easy enough to attribute the problem with solutions to the drive for profit. When I was initially conceptualizing this book, I taught a class called "The Problem with Solutions." In that class we discussed several of the readings I have cited throughout this book. By the end of the course we had made a list of about thirty problems with solutions, many of which exceeded issues with profit. Students noted the mismatch of solution to problems, the attraction of disruption for disruption's sake, the propensity for unintended consequences, the racial assumptions inhering in certain technologies, the performativity of solution-making, and more. And yet, time and time again, my students claimed that the problems we discussed were fundamentally about profit. I'm not going to completely disagree, but to see the problem with solutions as only cynical attempts at profit-making misses a lot that is important.

Consider Norman Borlaug, who "fathered" the Green Revolution with the development of high-yielding seed varieties. These seeds, along with the chemical inputs that nurtured them, turned out to be profitable for multinational corporations, but it is not at all clear that Borlaug set out to bolster profits. He firmly believed that improving output would reduce hunger. In the current context of solution-making, it happens that a lot of these products are not all that profitable, which is why many of the companies don't last. Moreover, many of the companies developing these solutions are B corporations—those that allow multiple bottom lines—and so they are not obligated to make profit the highest priority. For that matter, solutions (and solution-makers) are being made in institutions that are not strictly profit-making institutions, including the university, as well as the nonprofit sector and state regulatory agencies. Finally, many people developing these solutions are driven by genuine good intent, as Borlaug was, and may even be critical of what capitalism has wrought. But good intent doesn't necessarily make for good solutions, especially if they are guided by the impulses of the techno-fix (ignoring the underlying social causes), solutionism (letting the solution drive the problem), and the will to improve (satisfying the giver more than the receiver). These are all impulses that exceed profit making.

Rather, the problem with solutions is that they have been profoundly shaped by a way of thinking, replete with an ostensible optimism, a can-do spirit, and a great deal of hubris, that actually constrains the world of possibility by constraining our imaginations of what is possible. This way of thinking about and promoting change—a zeitgeist if you will—originated in Silicon Valley and has since permeated all manner of institutions, including the university, the nonprofit sector, and even government—after all, it is the government that often funds research into solutions. Although this way of thinking is not only about profits, I would be lying if I didn't acknowledge the ways that capitalism gave rise to it, profoundly shaping the conditions of solution-making. Recall that Silicon Valley rose to cultural and economic prominence following a major crisis in US capitalism. Seeing the potential of high-tech innovation to restore the country's economic standing, the government gave the tech sector virtual free rein, with favorable taxes and other incentives to accelerate its expansion.

The same notions of political economic governance (or neoliberalism) that gave Silicon Valley a huge injection of economic and moral support devalued public action and especially regulatory measures that threatened business profitability or incurred government costs. Instead, these new notions of governance insisted that the ensuing economic precarity, as well as intensified exploitation of natural resources, be addressed with entrepreneurship and other market-oriented approaches. Silicon Valley, with its start-up culture and win-win technologies promising both profits and social improvement, was well positioned to deliver on those terms. But it wasn't only the political economy that gave rise to the particular way of thinking. It was a culture of hype and innovation for innovation's sake. Here is where the "Californian ideology" comes in, a sensibility stemming from the state's contradictory political history that fused libertarianism and countercultural ideals and matched them to the unfettered optimism, opportunism, and boosterism marking its history (captured in the old saying "go west, young man").[4]

This way of thinking was also founded on and cultivated a kind of anti-intellectualism. This entailed a willful disregard of history, sociology, geography—remember, many of the techies were college dropouts—as well as an embrace of the shallow intellectualism of thought leadership, epitomized by TED Talks and other venues through which "big ideas" are communicated in small and simplistic doses.[5] Anti-intellectualism is central to the project of solutions. If you know a problem deeply, you realize that it can't be solved with a silver bullet. So not knowing helps absolve you of responsibility for the need for more substantive reform.[6] How optimistic is that?

So when Silicon Valley got bored with making faster computers, better gadgets, and disrupting taxicab businesses and hotels, and turned to the world's biggest challenges, it brought all that baggage. And it brought the conceit that it is possible to change capitalism through capitalism, a conceit that basically allowed those who have benefitted from capitalism to set the terms of engagement in social and ecological betterment. Writing on the allure of working within capitalism to change capitalism, investigative journalist Anand Giridharadas points out the folly of letting those who have succeeded define the terms of social change. Of course they deploy tools that created the problems. Of course they come up with solutions that do not challenge their wealth and status. Giving back charitably, as Giridharadas persuasively argues, is not the same as doing less harm and is especially pernicious when amassing wealth has come from doing harm.[7]

In short, in defining the terrain by which change should occur, the Silicon Valley model of solution-making has both delegitimized *and* made more difficult approaches that are not investible, deliverable, immediate, practical, technical, win-win, self-serving, or any of the other ways in which Silicon Valley as synecdoche goes about its work. Rendering social change as solutions, in other words, has obscured the real moonshot of changing the world to make it a more socially just, less violent, and ecologically livable place. How exactly is that optimistic? In Borlaug's day these politics were at least explicit: there were other possibilities on the table, most notably land reform to which the Green Revolution was conceived as an alternative. These days we can hardly see what might be otherwise—or how we might accomplish it.

Solutions, to be clear, are not intended to address the structural realities produced by capitalism, colonialism, racism, patriarchy, and so forth. They are a form of world-changing that purposefully avoids the frictions of social differences and political contentiousness. They in fact are a substitute for movement building, organizing, strategy, rebellion, prefigurative politics (referring to modeling the future you'd like to have), or any other action through which most progressive social change has been achieved. Don't take my word for it. Heed the words of Harvard Business School finance professor Mihir Desai, who in 2023 wrote:

> Many corporations have come to embrace broader social missions in response to the desire of younger investors and employees to use their capital and employment as instruments for social change. Another manifestation of magical thinking is believing that the best hopes for progress on our greatest challenges—climate change, racial injustice and economic inequality—are corporations and individual investment and consumption choices rather than political mobilization and our communities.[8]

If political mobilization is our best hope, why is it persistently cast as impossible while moonshot techno-solutions are championed? If critique allows us to imagine better worlds, why is it persistently cast as pessimistic while the magical thinking inhering in solutions gets a pass? If solutions are so limited, why do we think they can obviate despair?

## On Bringing Better Futures into Being and the Optimism of Critique

While working on this book, in a moment of serendipity I picked up *How to Do Nothing: Resisting the Attention Economy*, by artist and writer Jenny Odell, who had grown up in (and now teaches in the heart of) Silicon Valley. The book, it turned out, is not really about doing nothing.[9] It is about refusing to do the somethings we are constantly prevailed upon to do. It is about refusing productivity, efficiency, and the cult of individual entrepreneurship, rejecting "progress" and "disruption" as ideals, and carving out space and time for other ways of being, including engaging in slow acts of noticing. Paradoxically, I think Odell's book on doing nothing provides clues about what we should be doing. For one, it suggests a refusal of the idea that it is always better to do something: that we must act, urgently, or else we make things worse than before. Sometimes doing something can make things worse, if indeed that something leads to consequences far worse than what was there before.[10] And sometimes doing something overshadows better paths forward. Yet I think the primary lesson to draw from Odell is that we should refuse the imperative of solutions defined by Silicon Valley. We should refuse that question my father asked me when I was a college student about a solution to capitalism and instead ask ourselves how else to approach the manifold problems in the world that absolutely do require attention. And we should attend to these as problems and ask what they need, rather than jump to solutions.

A better approach, it seems to me, inverts the problems with solutions, an approach that I call response. Unlike solutionism, response suggests our actions should be informed by knowledge of how conditions came to be as they are. Before drumming up a solution, response asks us to do the intellectual work of first understanding a given problem, including its origin, the contingent way in which it has developed (not just capitalism, but how, in that place, at that time, by whom), and how and for whom it appears as a problem in the present. By asking what is needed, response by definition assumes an antecedent event or condition that warrants analysis. That makes a prefabricated solution, one absent a clear problem, nonsensical. Bringing it back to a concrete example explored in this book, a response would first explore the manifold reasons that farmers in a given place have come to rely on highly toxic pesticides in order to find ways to avert that tendency, rather than begin with a cool drone and look for a rationale to introduce it to farmers.

Unlike the techno-fix, response asks us to contend with the root social causes of a problem and consider how those might be addressed before resorting to a technical solution. As such, a response would recognize when vested interests benefit from perpetuating the problem or advocate for a techno-fix in order to subvert an approach that threatens their investments. A response wouldn't assume that problems can be solved without political contestation and ultimately a change in social worlds. Bringing it back to the concrete, a response would look at the central role of the conventional livestock industry in generating unsustainable and inhumane practices and might consider regulation of the industry or other economic incentives to thwart its practices, rather than simulating animal products that coexist with their animal-based counterparts and sometimes even under the same corporate umbrella.

Unlike the will to improve, response requires understanding who a problem affects, who speaks for the problem, and who *should* speak for the problem. It demands taking stock of our own positionality in relation to the problem and checking the need to do something on behalf of others if it's for our own emotional satisfaction. Unless we are directly harmed by the problem, it means finding ways to act as allies to those who are directly harmed, in respect of their needs and desires. In other words, a response doesn't imagine that those of us coming from places of geographic or situational privilege are better suited to solve a problem in the place it most manifests. To the contrary, it suggests we might be better positioned to redress how our literal place of privilege contributes to the problem elsewhere. In concrete terms again, it might mean working on our own country's food export policies to make them less likely to contribute to food insecurity globally rather than bringing an antipoverty scheme such as microlending or a moringa co-op to a location in the Global South.

Simply put, response asks for different sensibilities than those cultivated by Silicon Valley and replicated in the university where tomorrow's changemakers are incubated. While Silicon Valley indulges the performance of urgency, hubris, and can-do action, response cultivates qualities of patience, humility, and reflexivity: patience to recognize that meaningful change doesn't happen all at once, humility to assume that no one person has the answer and that it is almost always good to learn from others, and reflexivity to learn from past mistakes so that new attempts to address a problem avoid replicating failures from before.

None of this is to say that response is easy—and it can be downright daunting. That is where strategy comes in. Strategy is what bridges the long-term—and overwhelming—goals of societal transformation with the do-ability that makes solutions so attractive. It allows the problem to remain big and amorphous but the path forward to be specific. Strategy allows you to maintain and finesse the critique of such big problems as racial capitalism or the environmental destruction associated with capitalism, without basking in despair. For it is strategy that takes the immediate and particular manifestation of the larger phenomenon, the grandest challenge, and looks for fractures or vulnerabilities around which to organize, knowing the results may not be tangible or even evident for a time.[11] The evident successes hopefully make things marginally better, but even failure can be productive in contemplating new strategic directions. Engaging in strategy takes recognition that a big ship doesn't pivot easily and that it takes multiple efforts on multiple fronts to make things better. Rather than a *fix*, it is a *process* of amelioration.

The question then becomes how to identify the strategies that can lead to better worlds and not reinforce that which should be left behind. Here we can draw on the work of geographer Ruth Gilmore. While best known for scholarship and organizing around prison abolition, Gilmore's reflections on activism more generally has inspired those working in other areas as well. Building on the work of social philosopher Andre Gorz, Gilmore offers the notion of nonreformist reform, as distinct from reformist reform. Reformist reform lets the current order set the terms of how change can and should take place. This is what Silicon Valley does in spades: makes activism feel impossible while encouraging problem-solving that works well for Silicon Valley. In contrast, nonreformist reform, as described by Gilmore, is "deliberate change that does not create more obstacles in the larger struggle."[12] What she is offering is a standard not very daunting at all: work on things that move the ship in the right direction.

Thinking with Gilmore opens up entirely new fields of action, previously foreclosed when we look only for solutions. Nonreformist reform can be most anything that takes steps in more liberatory, just, and ecologically sound directions rather than reinforces the status quo. This can range from electoral campaigns to community organizing to mass mobilizations, from pressuring political representatives to outright resistance, from developing prefigurative alternatives to doing everything to undermine current systems. It might even include technological solutions if those are what is called for. Most critically, nonreformist reform rejects the notion that these more public- and community-facing approaches are so impossible as to be laughable. It simply won't do to believe in technological moonshots or the much lesser solutions that make up the lion's share of Silicon Valley's offerings, while conceding the ground on just about everything else.

I know this all sounds very abstract, and you're likely waiting for me to tell you exactly what actions to take or at least point you to exemplary programs or campaigns. I am hesitant to suggest that there are perfect situations in which to work, because there aren't. Organizations are messy, politics are fraught, tactics can be wrongheaded, strategies can fail. Moreover, fields of change are dynamic, and a part of strategy is to attend to that dynamism and find moments of vulnerability and responses to match. Examples I might highlight could be obsolete when you read this. Understanding the dynamism, frictions, and failures is an important part of the social change process from which better practices and strategies come. As but one example, that many environmental and food justice organizations now center BIPOC concerns and forward BIPOC leadership owes in large part to both activist and scholarly critique that past efforts marginalized the concerns and cultural idioms of BIPOC people. Critique, indeed, is an important component of praxis and critical for response and strategy. Implicit to the idea of critique is the imagination that things could be much better. This is why the Community Studies pedagogy doesn't shy away from critique.

So, in suggesting what you ought to do, I'd say start with being an informed citizen (yes, paying attention to the news), not shying away from critique, and voting. But mainly what I think you should do is resist the pull of solutions and embrace response instead, in an area of concern that moves you. Solutions won't absolve you—and they certainly won't save us from the crises of the present moment.

## Solutions Won't Save Us

When I sat down to write this book in late 2022, nearly five years had passed since I initially conceived of it. Besides the always daunting task of organizing my thoughts and writing in compelling prose, I faced another struggle. The years that had passed were the Trump years and their aftermath. In the context of all that had happened—the kids in cages at the border, the Muslim ban, the white supremacist march in Charlottesville, the perverse COVID response, more police killings of Black people, the election denialism, the insurrection at the US Capitol, the banning of books from school libraries, the retraction of reproductive choice, the violence against transgender people including denial of gender-affirming care, and yet more mass shootings—this book felt trivial, nitpicky, and perhaps not relevant for the present moment. The world felt on the verge of fascism, still does, and I didn't see what this book could say to that.

So I reflected on what the tech sector, bent on solutions, had done during those years. It had certainly grown more economically and culturally important while ever more absurd in its futuristic ambitions. In a few short years we saw the launch of the virtual universe of Meta and tech titans chartering their own private trips to the moon. We saw the advent of advanced AI and the very credible use of it for deepfakes. With its powerful social networks, the tech sector had, if anything, played a facilitating role in the election denialism and the US Capitol insurrection through unconstrained abuse of social media and disinformation campaigns. It had profited mightily through COVID-19, playing on newfound fears of personal contact to finally find a market for food-ordering apps and delivery robots, while doing very little about the ever more potent and evident problem of global warming. And it basically had nothing to say to the immanent threats to democracy, in the United States or elsewhere. To the contrary, it became increasingly clear that some of the biggest tech titans were self-avowed white nationalists or had far right political affiliations.[13] These were the guys investing in the future; these were the guys touting making the world a better place; these were they guys forging solutions. Why would we trust a system of innovation that these guys have made (and that has made these guys) to deliver our world out of its current mess?

What that also tells me is how empty solutions can be. No solutions as I have defined them in this book could reasonably reverse the manifold affronts to our bodies, lives, and already imperfect democracy itself. Indeed, there is no "solution" to fascism. A techno-fix won't stop it, neither algorithm, nor better bandwidth, nor cellular meat burger, nor vertical farm, nor app, nor advanced AI (especially not), nor anything remotely high tech, even if a moonshot. Dalliances with fascism demand a response—a political response. So maybe the very biggest problem with solutions is that in indulging our hopes for fixes, we lose the capacity for real problem-solving. Indeed, it can be difficult to imagine how to address problems that clearly have no techno-fix because we have created a culture in which the essential qualities for political struggle are no longer shared, supported, or taught in the institutions that shape our thinking. This culture and training warrants refusal.

# *Acknowledgments*

I am so very grateful for the many colleagues, students, and friends who encouraged me to get this book out of my system (a special shout-out to Aaron Bobrow-Strain who was first to entertain my idea and say "do it"). My original inspiration came at the tail end of my fellowship at the Radcliffe Institute in 2017–18. Some of those initial visits to the Boston food tech scene lit a fire. Thanks to Lisa Haushofer and Susanne Freidberg for joining me on those expeditions, and extra gratitude to Johanna Gilligan and Nicole Negowetti with whom I first discussed the idea for a book with a title of "the problems with solutions" and even developed a plan for writing together until we parted ways geographically and other work took precedence. I was also encouraged by audiences at places I presented on the problem with solutions: Tufts and Chatham Universities as well as the several universities I presented at on Silicon Valley agrifood tech research (Oxford, Pace, University of Kentucky, Temple, and the California Institute for Integral Studies in conversation with Larissa Zimberoff).

A grant from the National Science Foundation (Award # 1749184) to conduct more sustained research on the tech sector's aspirations to fix food and agriculture very much shaped the particular emphasis of this book. My collaborators on this research—Charlotte Biltekoff, Kathryn De Master, Madeleine Fairbairn, Zenia Kish, and Emily Reisman—enriched the intellectual work profoundly. I am no less thankful to one undergraduate and two graduate student researchers whose hands-on research supported the claims in this book: Anand Kumar, Summer Sullivan, and especially Michaelanne Butler, who worked tirelessly on data analysis and interviews. The AFTeR Project, as we came to call it, for Agri-Food Tech Research Project, has been a late-career intellectual joy.

This book was also influenced by my two decades of teaching in Community Studies at the University of California–Santa Cruz. Community Studies provides a model for engaged, social justice pedagogy that truly stands apart. Thanks to Mary Beth Pudup and Andrea Steiner for keeping the program alive, and the many students who come through the major inspired to do work much more profound than solutions. I have imagined you in many of the pages I wrote, thinking of the ideas that have excited you over the years. A special thanks to the students who took the chance on a class called "Problem with Solutions" in the early days of the pandemic. You guys rocked, and your enthusiasm for the material and pleas to write the book definitely helped make it happen.

A book title carried around in one's head for five years doesn't make a book. (And, by the way, thank you to Kimberly Thomas for not taking offense when I told her I was using a title that I hadn't known she had used for an article.) It takes sitting down and writing. I am so appreciative that I was able to return to Mesa Refuge as an alum and another writer's refuge I found on Whidbey Island, Washington. These retreats came at crucial times in the book's development, without which I couldn't have made such quick progress. A two-quarter sabbatical from UCSC helped too.

One of the best parts of writing, besides holing up in writers' cabins if that's your jam, is the opportunity to engage in conversation with brilliant and trusted colleagues. I owe utmost gratitude to four close colleagues who were generous enough to read, comment on, and workshop my first complete draft of the book: Charlotte Biltekoff, Aaron Bobrow-Strain, Madeleine Fairbairn, and Susanne Freidberg. Looking back, I realize that first draft was pretty raw, but I wasn't able to see how until I heard their incisive comments. Thank you.

Special thanks to Charlotte Biltekoff once more for being a sounding board and otherwise fantastic person to think with. Some of the ideas in the book I first introduced in journal articles, several written with Charlotte. These articles, too, had the benefit of constructive critiques not only from AFTeR Project members but a larger network of scholars working at the intersection of agrifood studies and science and technology studies. STSFAN, as we call it, is a rare bird in academia: an intentional, international intellectual community of scholars that meets regularly by Zoom to workshop papers and think together. I feel fortunate to have cocreated this group, also late in my career, and would like to spread the gospel of our "silencing the author" workshopping style that benefits both authors and commenters.

I'd be remiss if I did not acknowledge those who gave stellar feedback in the later stages of writing. Jessica Dempsey, Jesse Goldstein, and Mike Goodman really grasped what I was doing and provided lengthy and spot-on comments. I also appreciate several anonymous reviewers who both cheered and challenged me, and helped me to see what I really wanted to say. Thanks to Beth Clevinger and Sikina Jinnah whose efforts to draw me to another press was as uplifting as their comments were helpful. The awesome Kate Marshall at the University of California Press was a champion from the get-go but really stepped up in the final stages to strengthen the work with her ever-insightful and supportive ways.

I wish to extend my deepest thanks to *the* Sierra McCormick and her dad, Mike, too. How cool is it to have your adult child, with whom you struggled long ago over their writing, flag some of the errors and problem passages in the very first drafts of chapters and otherwise inspire you to up your writing game? Even Mike chipped on a couple of chapters when he wasn't doing the vast majority of household chores to make my writing possible. These two generous and kind people continue to be the ones who most closely and lovingly inhabit my life. Mike, I promise I'll stop working so hard soon. Really, I will. I'm looking forward to growing even older together. Sierra, I hope this book inspires change-making that will make the future more bearable than it seems it will be now.

# Glossary of Terms

APPROPRIATIONISM  The process of commodifying processes and inputs once produced on the farm and selling them back to farmers; leads to rural intensification.

CAPABILITIES APPROACH  A framework developed by Amartya Sen that focuses on the abilities individuals have to achieve the lives they want.

DESIGN THINKING  An approach to innovation that ascertains what users want and then ideates and tests solutions for feasibility and economic viability.

ECOMODERNISM  A school of environmental thought that advocates for technologies that allow humans to maintain (or reach) high standards of living.

ENDOWMENTS  In relation to the capabilities approach, an individual's original capabilities and resources to meet their needs.

ENTITLEMENTS  In relation to the capabilities approach, what an individual with a particular endowment can legally acquire to meet their needs.

GREEN REVOLUTION  Circa 1940–1970s, a massive effort by governments, foundations, and nonprofit organizations to increase food output in the developing world largely through breeding and promotion of high-yielding plant varieties, along with agrochemical inputs.

INTENSIFICATION  In regard to agriculture, using technology to increase output per unit of land.

MOONSHOTS  Technologies so radical and potentially transformative so as to appear near impossible.

NEOLIBERALISM  A theory of political economic governance that seeks to reduce the role of the state and allow markets to allocate goods and services; neoliberal governance also favors individual entrepreneurialism and personal responsibilities over dependence on states.

NEO-MALTHUSIAN  An analysis that attributes environmental problems to overpopulation in relation to resources.

NONDISRUPTIVE DISRUPTION  A term coined by Jesse Goldstein to denote the limits on technological creativity imposed by the requirements of capital for near-term profitability.

PROBLEMATIZATION  A term used by Tania Murray Li to denote the process of naming and characterizing a deficiency that needs rectification, connected to rendering technical.

PROBLEM CLOSURE  A term coined by Maarten Hajer to denote the tendency to define problems in relation to socially acceptable solutions.

PRODUCTIVISM  State support of an intensive, industrially-based, and expansionist agriculture oriented toward enhanced output and increased productivity.

RENDERING TECHNICAL  In relation to Tania Murray Li's notion of problematization, a process of bounding a problem to make it actionable with the tools of experts.

RESPONSES  In this book, an approach to social change that begins with analysis of what a situation needs.

RESPONSIBLE INNOVATION  An approach to technological problem-solving that attends in advance to potential consequences or public concerns as a technology is being developed.

SOLUTIONISM  A term popularized by Evgeny Morozov to denote when innovators presume rather than investigate the problems that they try to solve or develop a solution before defining a problem.

SOLUTIONS  In this book, finite, narrowly conceived fixes to problems that themselves have been bounded and rendered solvable.

SUBSTITUTIONISM  The process of shifting rural production to factories, made possible by the availability of cheaper or industrially produced raw materials; shifts value away from farm production into indoor or urban settings.

TECHNO-FIX  An approach using technological advances to solve societal problems, guided by the assumption that humans can engineer their way out of crises and that creating a technological solution is easier than changing societal structures or people's behaviors.

TECHNO-OPTIMISM  The view that science and technology properly applied can solve major society problems, closely related to the techno-fix.

WHITE SAVIOR COMPLEX  The neocolonial impulse of white people to improve or modernize the lives of others.

WILL TO IMPROVE  A phrase used by Tania Murray Li to refer to the impulse to act on behalf of others in order to better their lives, closely related to White Savior Complex.

# Notes

### Preface

1. Sarah Laskow, "A Distant Voyage, a Powerful Plant, and a Crowd-Backed Quest to Crack the Snack Market," *Fast Company*, April 24, 2014, www.fastcompany.com/3029485/the-amazing-plant-powering-a-quest-to-crack-the-crowded-snack-market; Michelle Paratore, "Kuli Kuli: The Next Superfood, and a Way to Support Women in West Africa," *Edible Startups*, December 4, 2013, https://ediblestartups.com/2013/12/04/kuli-kuli-the-next-superfood-and-a-way-to-support-women-in-west-africa; and Luke Tsai, "Kuli Kuli: Oakland Startup Touts West African 'Superfood,'" *East Bay Express*, December 10, 2013, www.eastbayexpress.com/WhatTheFork/archives/2013/12/10/kuli-kuli-oakland-startup-touts-west-african-superfood.

2. Louisa Burwood-Taylor, "Eighteen94 Capital Leads Funding in Kuli Kuli," *AgFunder News*, January 17, 2017, https://agfundernews.com/breaking-kelloggs-vc-1894-capital-makes-first-investment-agfunder-alum-kuli-kuli.html.

3. Kuli Kuli, "About Us," accessed January 2, 2019, www.kulikulifoods.com/about.

4. Melissa Leach, *Rainforest Relations* (Washington, DC: Smithsonian Institution Press, 1994); and Richard Schroeder, "Shady Practice—Gender and the Political Ecology of Resource Stabilization in Gambian Garden Orchards," *Economic Geography* 69, no. 4 (1993): 349–65.

5. Ananya Roy, "Governing Poverty," in *Encountering Poverty: Thinking and Acting in an Unequal World*, ed. Ananya Roy et al. (Oakland: University of California Press, 2016), 70.

6. Gregory Ferenstein, "Silicon Valley's New Politics of Optimism, Radical Idealism and Bizarre Loyalties," *The Guardian*, November 10, 2015, www.theguardian.com/us-news/2015/nov/10/silicon-valley-politics-tech-industry.

7. Sharon Stein et al., "Beyond Colonial Futurities in Climate Education," *Teaching in Higher Education* 28, no. 5 (2023): 987–1004.

## Introduction

1. Target, "Target Launches Collaboration with MIT's Media Lab and IDEO to Explore the Future of Food," news release, October 19, 2015, https://corporate.target.com/article/2015/10/mit-media-lab-collaboration; and LinYee Yuan, "MIT OpenAg Releases the Personal Food Computer 3.0, a Stem-Friendly Collaboration with Educators," *Mold*, October 24, 2018, https://thisismold.com/space/farm-systems/mit-openag-releases-the-personal-food-computer-3-0-a-stem-friendly-collaboration-with-educators#.W_MI6dMvzBL.

2. MIT Media Lab, "Launching the OpenAG Initiative at the MIT Media Lab," *Medium.com*, October 16, 2015, https://medium.com/mit-media-lab/launching-the-openag-initiative-at-the-mit-media-lab-7547f0d994a6.

3. Niyati Shah, "Open Agriculture Initiative: Is Digital Farming the Future of Food?," *Foodtank*, May 2016, https://foodtank.com/news/2016/05/open-agriculture-initiative-digital-farming.

4. Harry Goldstein, "MIT Media Lab's Food Computer Project Permanently Shut Down," *IEEE Spectrum*, May 17, 2020, https://spectrum.ieee.org/mit-media-lab-food-computer-project-shut-down.

5. Tom McKay, "MIT Built a Theranos for Plants," *Gizmodo*, September 8, 2019, https://gizmodo.com/mit-built-a-theranos-for-plants-1837968240.

6. Don Seiffert, "MIT Silent over Project Shutdown, Policies," *Boston Business Journal*, October 11, 2022, www.bizjournals.com/boston/news/2022/08/11/mit-silent-over-project-shutdown-policies.html; and Noam Cohen, "M.I.T. Media Lab, Already Rattled by the Epstein Scandal, Has a New Worry," *New York Times*, September 22, 2019, www.nytimes.com/2019/09/22/business/media/mit-media-lab-food-computer.html.

7. On making improvement strange, see Tania Murray Li, *The Will to Improve: Governmentality, Development, and the Practice of Politics* (Durham, NC: Duke University Press, 2007), 3.

8. Sean F. Johnston, "Alvin Weinberg and the Promotion of the Technological Fix," *Technology and Culture* 59, no. 3 (2018): 620–51.

9. Breakthrough Institute, "An Ecomodernist Manifesto," 2015, accessed December 23, 2019, https://ecomodernistmanifesto.squarespace.com.

10. Arne Naess, "The Shallow and the Deep, Long-Range Ecology Movement. A Summary," *Inquiry* 16 (1973): 95–100.

11. Naomi Klein, *This Changes Everything: Capitalism vs. the Climate* (New York: Simon and Schuster, 2015).

12. Naess, "Shallow and the Deep," 53. See also Michael Huesemann and Joyce Huesemann, *Techno-Fix: Why Technology Won't Save Us or the Environment* (Gabriola Island, BC: New Society Publishers, 2011); and Howard P. Segal, "Practical Utopias: America as Techno-Fix Nation," *Utopian Studies* 28, no. 2 (2017): 231–46.

13. Klein, *This Changes Everything*, 3; also Kari Norgaard, *Living in Denial: Climate Change, Emotions, and Everyday Life* (Cambridge, MA: MIT Press, 2011).

14. Ed Yong, "America Is Trapped in a Pandemic Spiral," *The Atlantic*, September 19, 2020, www.theatlantic.com/health/archive/2020/09/pandemic-intuition-nightmare-spiral-winter/616204.

15. Evgeny Morozov, *To Save Everything, Click Here: Technology, Solutionism, and the Urge to Fix Problems That Don't Exist* (London: Penguin, 2013), 6.

16. Morozov, *To Save Everything, Click Here*, 6.

17. Maarten A. Hajer, *The Politics of Environmental Discourse: Ecological Modernization and the Policy Process* (New York: Oxford University Press, 1995), 22.

18. David Wagner, *What's Love Got to Do with It? A Critical Look at American Charity* (New York: New Press, 2000).

19. Eric Wolf, *Europe and the People without History* (Berkeley: University of California Press, 1982); and Albert Memmi, *The Colonizer and the Colonized* (Boston: Orion Press, 1967).

20. Li, *Will to Improve*, 2–9. On how techno-solutions embed inventor values, see also Ruha Benjamin, *Race after Technology: Abolitionist Tools for the New Jim Code* (Cambridge, UK: Polity Press, 2019); Kelly Bronson, *The Immaculate Conception of Data: Agribusiness, Activists, and Their Shared Politics of the Future* (Montreal: McGill-Queen's University Press, 2022); and Sheila Jasanoff, *States of Knowledge: The Co-Production of Science and Social Order* (London: Routledge, 2004).

21. Li, *Will to Improve*, 7.

22. Li, *Will to Improve*, 2. See also Timothy Mitchell, *Rule of Experts: Egypt, Techno-Politics, Modernity* (Berkeley: University of California Press, 2002); and James Ferguson, *The Anti-Politics Machine: Development, Depoliticization, and Bureaucratic Power in Lesotho* (Minneapolis: University of Minnesota Press, 1994).

23. On race and colonialism, see Stuart Hall, "The West and the Rest: Discourse and Power," in *Formations of Modernity*, ed. Stuart Hall and Bram Gieben (Cambridge, UK: Polity Press, 1992); and Cedric Robinson, *Black Marxism: The Making of the Black Radical Tradition*, 3rd ed. (Chapel Hill: University of North Carolina, 2021).

24. Teju Cole, "The White-Savior Industrial Complex," *The Atlantic*, March 21, 2012, www.theatlantic.com/international/archive/2012/03/the-white-savior-industrial-complex/254843.

25. Jack Stilgoe, Richard Owen, and Phil Macnaghten, "Developing a Framework for Responsible Innovation," in *The Ethics of Nanotechnology, Geoengineering and Clean Energy*, ed. Andrew Maynard and Jack Stilgoe (London: Routledge, 2020).

26. Stilgoe, Owen, and Macnaghten, "Developing a Framework for Responsible Innovation"; and Laurens Klerkx and David Rose, "Dealing with the Game-Changing Technologies of Agriculture 4.0: How Do We Manage Diversity and Responsibility in Food System Transition Pathways?," *Global Food Security* 24 (2020), https://doi.org/10.1016/j.gfs.2019.100347.

27. Richard Buchanan, "Wicked Problems in Design Thinking," *Design Issues* 8, no. 2 (1992): 5–21; Tim Brown, "Design Thinking," *Harvard Business Review* (June 2008), https://hbr.org/2008/06/design-thinking; and Tim Brown and Barry Katz, "Change by Design," *Journal of Product Innovation Management* 28, no. 3 (2011): 381–83.

28. Lee Vinsel, "The Design Thinking Movement Is Absurd," *STS-News.medium.com*, November 26, 2018, https://sts-news.medium.com/the-design-thinking-movement-is-absurd-83df815b92ea.

29. Tim Seitz, "The 'Design Thinking' Delusion," *Jacobin*, October 16, 2018, https://jacobin.com/2018/10/design-thinking-innovation-consulting-politics.

30. Vinsel, "Design Thinking Movement Is Absurd."

31. Gilles Paquet, *The New Geo-Governance: A Baroque Approach* (Ottawa: University of Ottawa Press, 2005), 315.

32. For "market world," see Anand Giridharadas, *Winners Take All: The Elite Charade of Changing the World* (New York: Vintage Books, 2018).

33. Much of this is recounted in David Harvey, *A Brief History of Neoliberalism* (New York: Oxford University Press, 2005); and Geoff Mann, *Disassembly Required: A Field Guide to Actually Existing Capitalism* (Oakland, CA: AK Press, 2013).

34. Harvey, *Brief History of Neoliberalism*.

35. Katharyne Mitchell and Matthew Sparke, "The New Washington Consensus: Millennial Philanthropy and the Making of Global Market Subjects," *Antipode* 48, no. 3 (2016): 724–49.

36. Chris Rojek, "'Big Citizen' Celanthropy and Its Discontents," *International Journal of Cultural Studies* 17, no. 2 (2014): 127–41.

37. Raymond L. Bryant and Michael K. Goodman, "Consuming Narratives: The Political Ecology of 'Alternative' Consumption," *Transactions of the Institute of British Geographers* 29, no. 3 (2004): 344–66.

38. Mitchell and Sparke, "New Washington Consensus"; and Ananya Roy, *Poverty Capital: Microfinance and the Making of Development* (New York: Routledge, 2010).

39. Ananya Roy et al., *Encountering Poverty: Thinking and Acting in an Unequal World* (Oakland: University of California Press, 2016); and Sarah Besky, *The Darjeeling Distinction: Labor and Justice on Fair-Trade Tea Plantations in India* (Berkeley: University of California Press, 2013).

40. Jill Lindsey Harrison, *From the Inside Out: The Fight for Environmental Justice within Government Agencies* (Cambridge, MA: MIT Press, 2019).

## Chapter 1. Silicon Valley and the Urge to Make the World a Better Place

1. Morozov, *To Save Everything, Click Here*, vii.

2. Margaret O'Mara, *The Code: Silicon Valley and the Remaking of America* (New York: Penguin, 2020), 2.

3. AnnaLee Saxenian, *Regional Advantage: Culture and Competition in Silicon Valley and Route 128* (Cambridge, MA: Harvard University Press, 1996).

4. Saxenian, *Regional Advantage*; O'Mara, *The Code*; and Tom Nicholas, *VC: An American History* (Cambridge, MA: Harvard University Press, 2019).

5. Marilynn Johnson, *The Second Gold Rush: Oakland and the East Bay in World War II* (Berkeley: University of California, 1994).

6. Fred Turner, *From Counterculture to Cyberculture: Stewart Brand, the Whole Earth Network, and the Rise of Digital Utopianism* (Chicago: Chicago University Press, 2006), 4.

7. Turner, *From Counterculture to Cyberculture*, 6.

8. Saxenian, *Regional Advantage*; O'Mara, *The Code*; and Nicholas, *VC: An American History*.

9. Saxenian, *Regional Advantage*, 38.

10. Saxenian, *Regional Advantage*.

11. O'Mara, *The Code*.

12. Malcom Harris, *Palo Alto: A History of California, Capitalism, and the World* (New York: Little, Brown & Co., 2023).

13. O'Mara, *The Code*.

14. Harris, *Palo Alto*.

15. Mariana Mazzucato, *The Entrepreneurial State: Debunking Public vs. Private Sector Myths* (New York: Public Affairs, 2015).

16. Nicholas, *VC: An American History*, 238.

17. Nicholas, *VC: An American History*, 237.

18. O'Mara, *The Code*, 73.

19. O'Mara, *The Code*, 73; Nicholas, *VC: An American History*; and Saxenian, *Regional Advantage*.

20. Fred Block and Matthew R. Keller, "Where Do Innovations Come From? Transformations in the US Economy, 1970–2006," *Socio-Economic Review* 7, no. 3 (2009): 459–83.

21. Nicholas, *VC: An American History*, 268.

22. O'Mara, *The Code*.

23. Nicholas, *VC: An American History*.

24. Marc Andreessen, "Why Software Is Eating the World," *Wall Street Journal*, August 20, 2011, www.wsj.com/articles/SB10001424053111903480904576512250915629460.

25. Susi Geiger, "Silicon Valley, Disruption, and the End of Uncertainty," *Journal of Cultural Economy* 13, no. 2 (2020): 169–84; and Jill Lepore, "The Disruption Machine," *The New Yorker*, June 16, 2014, 30–36.

26. Clayton M. Christensen, Michael E. Raynor, and Rory McDonald, "What Is Disruptive Innovation?," *Harvard Business Review* (December 2015), https://hbr.org/2015/12/what-is-disruptive-innovation.

27. Francis Jervis, "Eating the World: Iterative Capital after Silicon Valley" (PhD diss., New York University, 2020); and Turner, *From Counterculture to Cyberculture*.

28. Richard Barbrook and Andy Cameron, "Californian Ideology: A Critique of West Coast Cyber-Libertarianism," *Science as Culture* 6, no. 1 (1996): 44–72.

29. Stuart Hogarth, "Valley of the Unicorns: Consumer Genomics, Venture Capital and Digital Disruption," *New Genetics and Society* 36, no. 3 (2017): 250–72.

30. Eric Gianella, "Morality and the Idea of Progress in Silicon Valley," *Berkeley Journal of Sociology* (January 14, 2015), http://berkeleyjournal.org/2015/01/morality-and-the-idea-of-progress-in-silicon-valley; and Morozov, *To Save Everything, Click Here*.

31. Anastasia Rose O'Rourke, "The Emergence of Cleantech" (PhD diss., Yale University, 2009).

32. O'Mara, *The Code*, 296; and Jesse Goldstein, *Planetary Improvement: Cleantech Entrepreneurship and the Contradictions of Green Capitalism* (Cambridge, MA: MIT Press, 2018), 95.

33. Juliet Eilperin, "Why the Clean Tech Boom Went Bust," *Wired*, January 20, 2012, www.wired.com/2012/01/ff_solyndra.

34. Nick J. Fox, "Green Capitalism, Climate Change and the Technological Fix: A More-Than-Human Assessment," *Sociological Review* 71, no. 5 (2022), https://doi.org/10.1177/00380261221121232.

35. Jane Reisman, Veronica Olazabal, and Shawna Hoffman, "Putting the 'Impact' in Impact Investing: The Rising Demand for Data and Evidence of Social Outcomes," *American Journal of Evaluation* 39, no. 3 (2018): 389–95; and Cathy Clark, Jed Emerson, and Ben Thornley, *Collaborative Capitalism and the Rise of Impact Investing* (San Francisco: Jossey-Bass, 2014).

36. Rob Wyse, "Impact Investing Defined," *HuffPost*, August 30, 2011, www.huffpost.com/entry/impact-investing-defined_b_941916.

37. Madeleine Fairbairn and Emily Reisman, "The Incumbent Advantage: Corporate Power in Agri-Food Tech," *Journal of Peasant Studies* (2024).

38. Susan Cohen and Yael V. Hochberg, "Accelerating Startups: The Seed Accelerator Phenomenon," *SSRN Electronic Journal* (March 30, 2014), https://doi.org/10.2139/ssrn.2418000.

39. Nicholas, *VC: An American History.*
40. Giridharadas, *Winners Take All.*
41. Benjamin Wurgaft, *Meat Planet: Artificial Flesh and the Future of Food* (Oakland: University of California Press, 2019), 107; and Vinsel, "Design Thinking Movement Is Absurd."
42. Donald MacKenzie, "Is Economics Performative? Option Theory and the Construction of Derivatives Markets," *Journal of the History of Economic Thought* 28, no. 1 (2006): 29–55; and Anna Tsing, "Inside the Economy of Appearances," *Public Culture* 12, no. 1 (2000): 115–44.
43. Morozov, *To Save Everything, Click Here.*
44. Giridharadas, *Winners Take All.*
45. Goldstein, *Planetary Improvement.*

## Chapter 2. Agrifood Solutions before Silicon Valley

1. Gregg Easterbrook, "Forgotten Benefactor of Humanity," *The Atlantic*, January 1997, www.theatlantic.com/magazine/archive/1997/01/forgotten-benefactor-of-humanity/306101.
2. Norman Borlaug, "The Nobel Prize: Norman Borlaug Acceptance Speech," December 10, 1970, www.nobelprize.org/prizes/peace/1970/borlaug/acceptance-speech.
3. Thomas Robert Malthus, *An Essay on the Principle of Population*, ed. Geoffrey Gilbert (1798; Oxford, UK: Oxford University Press, 1999).
4. Wallace as cited in Eric B. Ross, *The Malthus Factor: Poverty, Politics, and Population in Capitalist Development* (London: Zed Books, 1998), 165.
5. Ross, *Malthus Factor*; and Joseph Cotter, *Troubled Harvest: Agronomy and Revolution in Mexico, 1880–2002* (Westport, CT: Greenwood Publishing Group, 2003).
6. Ross, *Malthus Factor*; and Cotter, *Troubled Harvest.*
7. Prabhu L. Pingali, "Green Revolution: Impacts, Limits, and the Path Ahead," *Proceedings of the National Academy of Sciences* 109, no. 31 (2012): 12302–8.
8. M. Wik, P. Pingali, and S. Broca, "Background Paper for the World Development Report 2008: Global Agricultural Performance: Past Trends and Future Prospects," World Bank, Washington, DC, 2008.
9. Pingali, "Green Revolution."

10. Scott Kilman and Roger Thurow, "Father of 'Green Revolution' Dies," *Wall Street Journal*, September 13, 2009, www.wsj.com/articles/SB125281643150406425.

11. Easterbrook, "Forgotten Benefactor of Humanity."

12. John H. Perkins, "The Rockefeller Foundation and the Green Revolution, 1941–1956," *Agriculture and Human Values* 7, no. 3 (1990): 6–18.

13. Nick Cullather, *The Hungry World: America's Cold War Battle against Poverty in Asia* (Cambridge, MA: Harvard University Press, 2010); Perkins, "Rockefeller Foundation and the Green Revolution"; and Ross, *Malthus Factor*.

14. Amartya Sen, *Poverty and Famines: An Essay on Entitlement and Deprivation* (New York: Oxford University Press, 1982), also in what follows.

15. Alex De Waal, *Mass Starvation: The History and Future of Famine* (Hoboken, NJ: John Wiley & Sons, 2017); and Stephen Devereux, *The New Famines: Why Famines Persist in an Era of Globalization* (London: Routledge, 2006).

16. Stephen Devereux, "Sen's Entitlement Approach: Critiques and Counter-Critiques," *Oxford Development Studies* 29, no. 3 (2001): 245–63.

17. Philip Lowe et al., "Regulating the New Rural Spaces: The Uneven Development of Land," *Journal of Rural Studies* 9, no. 3 (1993): 221.

18. Janet Poppendieck, *Breadlines Knee Deep in Wheat: Food Assistance in the Great Depression* (New Brunswick, NJ: Rutgers University Press, 1986).

19. Bill Winders, *The Politics of Food Supply: U.S. Agricultural Policy in the World Economy* (New Haven, CT: Yale University Press, 2009).

20. William Boyd and Michael J. Watts, "Agro-Industrial Just-in-Time: The Chicken Industry and Postwar American Capitalism," in *Globalising Food: Agrarian Questions and Global Restructuring*, ed. David Goodman and Michael J. Watts (London: Routledge, 1997).

21. Willard W. Cochrane, *The Development of American Agriculture* (Minneapolis: University of Minnesota Press, 1993).

22. Cochrane, *Development of American Agriculture*; and Philip H. Howard, *Concentration and Power in the Food System: Who Controls What We Eat?* (London: Bloomsbury Publishing, 2016).

23. Tarique Niazi, "Rural Poverty and the Green Revolution: The Lessons from Pakistan," *Journal of Peasant Studies* 31, no. 2 (2004): 242–60; Perkins, "Rockefeller Foundation and the Green Revolution"; and Raj Patel, "The Long Green Revolution," *Journal of Peasant Studies* 40, no. 1 (2013): 1–63.

24. Charles Mann, "Can Planet Earth Feed 10 Billion People?," *The Atlantic*, March 2018, www.theatlantic.com/magazine/archive/2018/03/charles-mann-can-planet-earth-feed-10-billion-people/550928.

## Chapter 3. Silicon Valley Bites Off Agriculture and Food

1. Alexia Tsotsis, "Monsanto Buys Weather Big Data Company Climate Corporation for around $1.1b," *TechCrunch*, October 2, 2013, https://techcrunch.com/2013/10/02/monsanto-acquires-weather-big-data-company-climate-corporation-for-930m/?guccounter=1.

2. Michael Kearney, "Cleantech's Comeback," *The Engine*, November 9, 2020, https://medium.com/tough-tech/cleantechs-comeback-38ec13507452; Eilperin, "Why the Clean Tech Boom Went Bust"; and Goldstein, *Planetary Improvement*.

3. Scott McFettridge, "The Weird State of Organic Farming in the U.S.: Consumers Love It, but Fewer and Fewer Farmers Are Converting," *Fortune*, September 22, 2022, https://fortune.com/2022/09/22/organic-farming-popular-but-not-with-farmers-converting.

4. AgFunder, "AgFunder AgriFoodTech Investment Report," 2022, accessed December 19, 2022, https://agfunder.com/research/2022-agfunder-agrifoodtech-investment-report.

5. Ben Fine, "Towards a Political Economy of Food," *Review of International Political Economy* 1, no. 3 (1994): 519–45; and David Goodman and Michael Redclift, "Constructing a Political Economy of Food," *Review of International Political Economy* 1, no. 3 (1994): 547–52.

6. Margaret FitzSimmons and David Goodman, "Incorporating Nature: Environmental Narratives and the Reproduction of Food," in *Remaking Reality: Nature at the Millenium*, ed. Bruce Braun and Noel Castree (London: Routledge, 1998); David Goodman, Bernardo Sorj, and John Wilkinson, *From Farming to Biotechnology* (Oxford: Basil Blackwell, 1987); Karl Kautsky, *The Agrarian Question* (1899; London: Zwan Press, 1988); and Susan A. Mann, *Agrarian Capitalism in Theory and Practice* (Chapel Hill: University of North Carolina Press, 1990).

7. Goodman, Sorj, and Wilkinson, *From Farming to Biotechnology*.

8. Goodman, Sorj, and Wilkinson, *From Farming to Biotechnology*; and Jack Kloppenburg, *First the Seed: The Political Economy of Plant Biotechnology* (Madison: University of Wisconsin Press, 2005).

9. Gyorgy Scrinis and Carlos Augusto Monteiro, "Ultra-Processed Foods and the Limits of Product Reformulation," *Public Health Nutrition* 21, no. 1 (2018): 247–52.

10. Frederick H. Buttel and Lawrence Busch, "The Public Agricultural Research System at the Crossroads," *Agricultural History* 62, no. 2 (1988): 303–24; Christopher R. Henke, *Cultivating Science, Harvesting Power* (Cambridge, MA: MIT Press, 2008); and Alan P. Rudy et al., *Universities in the Age of Corporate Science: The UC Berkeley-Novartis Controversy* (Philadelphia: Temple University Press, 2007).

11. Henke, *Cultivating Science, Harvesting Power*.

12. William Friedland, "The End of Rural Society and the Future of Rural Sociology," paper prepared for the Annual Meeting of the Rural Sociological Society, Guelph, Ontario, 1981.

13. Lawrence Busch et al., *Plants, Power, and Profit* (Cambridge, MA: Blackwell, 1991); Leland L. Glenna et al., "University Administrators, Agricultural Biotechnology, and Academic Capitalism: Defining the Public Good to Promote University-Industry Relationships," *Sociological Quarterly* 48, no. 1 (2007): 141–67; Steven A. Wolf and Frederick H. Buttel, "The Political Economy of Precision Farming," *American Journal of Agricultural Economics* 78, no. 5 (1996): 1269–74; and Rudy et al., *Universities in the Age of Corporate Science*.

14. Howard, *Concentration and Power in the Food System*.

15. Fairbairn and Reisman, "Incumbent Advantage."

16. AgFunder, "AgFunder AgriFoodTech Investment Report."

17. Gianella, "Morality and the Idea of Progress in Silicon Valley."

Chapter 4. Alternative Protein and the Nothing Burger of the Techno-Fix

This chapter draws on material first published in Charlotte Biltekoff and Julie Guthman, "Conscious, Complacent, Fearful: Agri-Food Tech's Market-Making Public Imaginaries," *Science as Culture* 32, no. 1 (2023): 58–82; Julie Guthman and Charlotte Biltekoff, "Agri-Food Tech's Building Block: Narrating Protein, Agnostic of Source, in the Face of Crisis," *BioSocieties* 18 (2023): 656–78; and Julie Guthman and Charlotte Biltekoff, "Magical Disruption? Alternative Protein and the Promise of De-Materialization," *Environment and Planning E: Nature and Space* 4, no. 4 (2021): 1583–600.

1. Maia Keerie, "Record $3.1 Billion Invested in Alt Proteins in 2020 Signals Growing Market Momentum for Sustainable Proteins," Good Food Institute, March 18, 2021, https://gfi.org/blog/2020-state-of-the-industry-highlights.

2. Warren J. Belasco, *Appetite for Change* (New York: Pantheon, 1989); Charlotte Biltekoff, *Eating Right in America: The Cultural Politics of Food and Health* (Durham, NC: Duke University Press, 2013); and Harvey A. Levenstein, *Paradox of Plenty: A Social History of Eating in Modern America* (New York: Oxford University Press, 1993).

3. Kenneth J. Carpenter, "The History of Enthusiasm for Protein," *Journal of Nutrition* 116, no. 7 (1986): 1364-70.

4. Aya Hirata Kimura, *Hidden Hunger: Gender and the Politics of Smarter Foods* (Ithaca, NY: Cornell University Press, 2013), 21.

5. Henning Steinfeld et al., "Livestock's Long Shadow: Environmental Issues and Options," Food and Agriculture Organization, 2006, accessed March 23, 2023, www.fao.org/3/a0701e/a0701e00.htm; and Tony Weis, *The Ecological Hoofprint: The Global Burden of Industrial Livestock* (London: Zed Books, 2013).

6. Hannah Landecker, "Antibiotic Resistance and the Biology of History," *Body & Society* 22, no. 4 (2015): 1-34.

7. Alex Blanchette, *Porkopolis: American Animality, Standardized Life, and the Factory Farm* (Durham, NC: Duke University Press, 2020).

8. Steve Hinchliffe et al., *Pathological Lives: Disease, Space and Biopolitics* (Chichester, UK: John Wiley & Sons, 2016); Blanchette, *Porkopolis*; and Marianne Elisabeth Lien, *Becoming Salmon: Aquaculture and the Domestication of a Fish* (Oakland: University of California Press, 2015).

9. Alexandra E. Sexton, Tara Garnett, and Jamie Lorimer, "Framing the Future of Food: The Contested Promises of Alternative Proteins," *Environment and Planning E: Nature and Space* 2, no. 1 (2019): 47-72; Erik Jönsson, Tobias Linné, and Ally McCrow-Young, "Many Meats and Many Milks? The Ontological Politics of a Proposed Post-Animal Revolution," *Science as Culture* 28, no. 1 (2019): 70-97; Michael J. Mouat and Russell Prince, "Cultured Meat and Cowless Milk: On Making Markets for Animal-Free Food," *Journal of Cultural Economy* 11, no. 4 (2018): 315-29; and Russell Hedberg, "Bad Animals, Techno-Fixes, and the Environmental Narratives of Alternative Protein," *Frontiers in Sustainable Food Systems* 7 (2023), https://doi.org/10.3389/fsufs.2023.1160458.

10. Kim Severson, "The New Secret Chicken Recipe? Animal Cells," *New York Times*, February 15, 2022, www.nytimes.com/2022/02/15/dining/cell-cultured-meat.html.

11. Hedberg, "Bad Animals, Techno-Fixes, and the Environmental Narratives of Alternative Protein."

12. Kimura, *Hidden Hunger*, 24–26.

13. Carpenter, "History of Enthusiasm for Protein," 1367.

14. Kimura, *Hidden Hunger*.

15. Blanchette, *Porkopolis*; and Steve Striffler, *Chicken: The Dangerous Transformation of America's Favorite Food* (New Haven, CT: Yale University Press, 2005).

16. Wurgaft, *Meat Planet*; Sexton, Garnett, and Lorimer, "Framing the Future of Food"; and Jönsson, Linné, and McCrow-Young, "Many Meats and Many Milks?"

17. Sexton, Garnett, and Lorimer, "Framing the Future of Food"; and Jönsson, Linné, and McCrow-Young, "Many Meats and Many Milks?"

18. Harriet Friedmann, "Distance and Durability: Shaky Foundations of the World Food Economy," in *The Global Restructuring of Agro-Food Systems*, ed. Philip McMichael (Ithaca, NY: Cornell University Press, 1994); and Jack Goody, "Industrial Food: Towards the Development of a World Cuisine," in *Food and Culture: A Reader*, ed. Carole Counihan and Penny Van Esterik (New York: Routledge, 2012).

19. Goodman, Sorj, and Wilkinson, *From Farming to Biotechnology*.

20. Goodman, Sorj, and Wilkinson, *From Farming to Biotechnology*.

21. Kregg Hetherington, *The Government of Beans: Regulating Life in the Age of Monocrops* (Durham, NC: Duke University Press, 2020).

22. Frances Moore Lappé, *Diet for a Small Planet* (New York: Ballantine Books, 1971).

23. A.C.Y. Lam et al., "Pea Protein Isolates: Structure, Extraction, and Functionality," *Food Reviews International* 34, no. 2 (2018): 126–47.

24. Dustin Mulvaney, *Solar Power: Innovation, Sustainability, and Environmental Justice* (Oakland: University of California Press, 2019).

25. Hedberg, "Bad Animals, Techno-Fixes, and the Environmental Narratives of Alternative Protein"; John Lynch and Raymond Pierrehumbert, "Climate Impacts of Cultured Meat and Beef Cattle," *Frontiers in Sustainable Food Systems* 3 (2019), https://doi.org/10.3389/fsufs.2019.00005; and Carolyn S.

Mattick et al., "Anticipatory Life Cycle Analysis of In Vitro Biomass Cultivation for Cultured Meat Production in the United States," *Environmental Science & Technology* 49, no. 19 (2015): 11941–49.

26. Hedberg, "Bad Animals, Techno-Fixes, and the Environmental Narratives of Alternative Protein."

27. Blanchette, *Porkopolis*.

28. Max Liboiron, "Modern Waste Is an Economic Strategy," *Lo Squaderno: Explorations in Space and Society* (special edition on Garbage and Wastes) 29 (2013): 9–12.

29. Erik Jönsson, "Benevolent Technotopias and Hitherto Unimaginable Meats: Tracing the Promises of in Vitro Meat," *Social Studies of Science* 46, no. 5 (2016): 725–48; and Wurgaft, *Meat Planet*.

30. Julie Guthman, "Binging and Purging: Agrofood Capitalism and the Body as Socioecological Fix," *Environment and Planning A* 47, no. 12 (2015): 2522–36.

31. Fairbairn and Reisman, "Incumbent Advantage."

32. Philip H. Howard et al., "'Protein' Industry Convergence and Its Implications for Resilient and Equitable Food Systems," *Frontiers in Sustainable Food Systems* 5 (2021), https://doi.org/10.3389/fsufs.2021.684181.

33. Klein, *This Changes Everything*.

34. Hedberg, "Bad Animals, Techno-Fixes, and the Environmental Narratives of Alternative Protein"; and Richard Teague and Urs Kreuter, "Managing Grazing to Restore Soil Health, Ecosystem Function, and Ecosystem Services," *Frontiers in Sustainable Food Systems* 4 (2020), https://doi.org/10.3389/fsufs.2020.534187.

35. Marina Bolotnikova, "Five Things to Know about the SCOTUS Challenge to California's Ban on Extreme Farm Animal Confinement," *The Counter*, March 29, 2022, https://thecounter.org/california-proposition-12-animal-welfare-supreme-court; and Frederick H. Buttel, "The Recombinant BGH Controversy in the United States: Toward a New Consumption Politics of Food?," *Agriculture and Human Values* 17, no. 1 (2000): 5–20.

36. Robert H. Smith, "Livestock Production: The Unsustainable Environmental and Economic Effects of an Industry Out of Control," *Buffalo International Law Journal* 4, no. 1 (1996): 45–129.

37. Klein, *This Changes Everything*.

38. Julie Creswell, "Beyond Meat Is Struggling, and the Plant-Based Meat Industry Worries," *New York Times*, November 21, 2022, www.nytimes.com/2022/11/21/business/beyond-meat-industry.html.

39. Creswell, "Beyond Meat Is Struggling"; and Laura Reiley, "Alt-Meat Fever Has Cooled. Here's Why," *Washington Post*, November 12, 2022, www.washingtonpost.com/business/2022/11/12/plant-based-meat-market.

## Chapter 5. Digital Technologies and Plowing Through to the Problem

Portions of this chapter were first published by Springer Nature in Julie Guthman and Michaelanne Butler, "Fixing Food with a Limited Menu: On (Digital) Solutionism in the Agri-Food Tech Sector," *Agriculture and Human Values* 40, no. 3 (2023): 835–48.

1. Busch et al., *Plants, Power, and Profit*.
2. Wolf and Buttel, "Political Economy of Precision Farming."
3. Emily Reisman, "Sanitizing Agri-Food Tech: Covid-19 and the Politics of Expectation," *Journal of Peasant Studies* 48, no. 5 (2021): 910–33; and Emily Duncan et al., "New but for Whom? Discourses of Innovation in Precision Agriculture," *Agriculture and Human Values* 38 (2021): 1181–99.
4. Duncan et al., "New but for Whom?," 1194; and Evagelos D. Lioutas and Chrysanthi Charatsari, "Big Data in Agriculture: Does the New Oil Lead to Sustainability?," *Geoforum* 109 (2020): 1–3.
5. Duncan et al., "New but for Whom?"; Lioutas and Charatsari, "Big Data in Agriculture"; Kelly Bronson, "Looking through a Responsible Innovation Lens at Uneven Engagements with Digital Farming," *NJAS: Wageningen Journal of Life Sciences* 90-91, no. 1 (2019): 1–6; Steven A. Wolf and Spencer D. Wood, "Precision Farming: Environmental Legitimation, Commodification of Information, and Industrial Coordination," *Rural Sociology* 62, no. 2 (1997): 180–206; and Oane Visser, Sarah Ruth Sippel, and Louis Thiemann, "Imprecision Farming? Examining the (in)Accuracy and Risks of Digital Agriculture," *Journal of Rural Studies* 86 (2021): 623–32.
6. Louisa Prause, Sarah Hackfort, and Margit Lindgren, "Digitalization and the Third Food Regime," *Agriculture and Human Values* 38, no. 3 (2021): 641–55.
7. Bronson, *Immaculate Conception of Data*, 31.

8. Bronson, *Immaculate Conception of Data*, 31.

9. Kelly Bronson and Irena Knezevic, "Big Data in Food and Agriculture," *Big Data & Society* 3, no. 1 (2016), https://doi.org/10.1177/2053951716648174; Alistair Fraser, "Land Grab/Data Grab: Precision Agriculture and Its New Horizons," *Journal of Peasant Studies* 46, no. 5 (2019): 893–912; and Wolf and Wood, "Precision Farming."

10. Benjamin, *Race after Technology*; Bronson, *Immaculate Conception of Data*; Shannon Elizabeth Bell, Alicia Hullinger, and Lilian Brislen, "Manipulated Masculinities: Agribusiness, Deskilling, and the Rise of the Businessman-Farmer in the United States," *Rural Sociology* 80, no. 3 (2015): 285–313; Bronson and Knezevic, "Big Data in Food and Agriculture"; Fraser, "Land Grab/Data Grab"; and Madeleine Fairbairn and Zenia Kish, "'A Poverty of Data'? Exporting the Digital Revolution to Farmers in the Global South," in *The Nature of Data: Infrastructures, Environments, Politics*, ed. Jenny Goldstein and Eric Nost (Lincoln: University of Nebraska Press, 2022).

11. Bronson and Knezevic, "Big Data in Food and Agriculture"; and Visser, Sippel, and Thiemann, "Imprecision Farming?"

12. Visser, Sippel, and Thiemann, "Imprecision Farming?"

13. Wolf and Wood, "Precision Farming."

14. Morozov, *To Save Everything, Click Here*, 6.

15. Susanne Freidberg, "Big Food and Little Data: The Slow Harvest of Corporate Food Supply Chain Sustainability Initiatives," *Annals of the American Association of Geographers* 107, no. 6 (2017): 1389–406.

16. AgThentic, "How Silicon Valley Set Agtech Back a Decade," *AgThentic Blog*, July 6, 2021, https://blog.agthentic.com/how-silicon-valley-set-agtech-back-a-decade-b9d46e0acfod.

17. Jamie Lorimer, "Probiotic Environmentalities: Rewilding with Wolves and Worms," *Theory, Culture & Society* 34, no. 4 (2017); and Julie Guthman, *Wilted: Pathogens, Chemicals, and the Fragile Future of the Strawberry Industry* (Oakland: University of California Press, 2019).

## Chapter 6. Silicon Valley Thinking Comes to the University

1. Claire Kremen, Alastair Iles, and Christopher Bacon, "Diversified Farming Systems: An Agroecological, Systems-Based Alternative to Modern Industrial Agriculture," *Ecology and Society* 17, no. 4 (2012), http://dx.doi.org/10.5751

/ES-05103-170444; and Marney E. Isaac et al., "Agroecology in Canada: Towards an Integration of Agroecological Practice, Movement, and Science," *Sustainability* 10, no. 9 (2018), https://doi.org/10.3390/su10093299.

2. Les Levidow, Michel Pimbert, and Gaetan Vanloqueren, "Agroecological Research: Conforming—or Transforming the Dominant Agro-Food Regime?," *Agroecology and Sustainable Food Systems* 38, no. 10 (2014): 1127–55.

3. Nicholas, *VC: An American History*, 185.

4. Nicholas, *VC: An American History*, 186.

5. Rudy et al., *Universities in the Age of Corporate Science*.

6. Elizabeth Popp Berman, *Creating the Market University: How Academic Science Became an Economic Engine* (Princeton, NJ: Princeton University Press, 2012).

7. Vinsel, "Design Thinking Movement Is Absurd"; Berman, *Creating the Market University*; Philip Mirowski, *Science-Mart: Privatizing American Science* (Cambridge, MA: Harvard University Press, 2011); and Lawrence Busch, *Knowledge for Sale: The Neoliberal Takeover of Higher Education* (Cambridge, MA: MIT Press, 2017).

8. Mirowski, *Science-Mart*, 339.

9. Busch, *Knowledge for Sale*.

10. Carl Rhodes, Christopher Wright, and Alison Pullen, "Changing the World? The Politics of Activism and Impact in the Neoliberal University," *Organization* 25, no. 1 (2018): 139–47; Andrew Gunn and Michael Mintrom, "Evaluating the Non-Academic Impact of Academic Research: Design Considerations," *Journal of Higher Education Policy and Management* 39, no. 1 (2017): 20–30; and Steven Hill, "Assessing (for) Impact: Future Assessment of the Societal Impact of Research," *Palgrave Communications* 2, no. 1 (2016): 1–7, 2.

11. C. Orr, "Academic Activism," in *The Women's Movement Today: An Encyclopedia of Third Wave Feminism*, ed. Leslie L. Heywood (Westport, CT: Greenwood, 2006).

12. Simon Fielke et al., "A Call to Expand Disciplinary Boundaries So That Social Scientific Imagination and Practice Are Central to Quests for 'Responsible' Digital Agri-Food Innovation," *Sociologia Ruralis* 62, no. 2 (2022): 151–61.

13. Jerry A. Jacobs and Scott Frickel, "Interdisciplinarity: A Critical Assessment," *Annual Review of Sociology* 35, no. 1 (2009): 43–65.

14. Karly Burch et al., "Social Science—STEM Collaborations in Agriculture, Food and Beyond: An STSFAN Manifesto," *Agriculture and Human Values* 40, no. 3 (2023): 939-49.

15. Garrett M. Broad, "Improving the Agri-Food Biotechnology Conversation: Bridging Science Communication with Science and Technology Studies," *Agriculture and Human Values* 40, no. 3 (2023): 929-38; Kelly Bronson, "Excluding 'Anti-Biotech' Activists from Canadian Agri-Food Policy Making: Ethical Implications of the Deficit Model of Science Communication," in *Ethics and Practice in Science Communication*, ed. Susanna Priest, Jean Goodwin, and Michael F. Dahlstrom (Chicago: University of Chicago Press, 2018); and Rachel Schurman and William Munro, *Fighting for the Future of Food: Activists versus Agribusiness in the Struggle over Biotechnology* (Minneapolis: University of Minnesota Press, 2010).

16. Isabelle Stengers, *Another Science Is Possible: A Manifesto for Slow Science* (Cambridge, UK: Polity Press, 2018).

17. Mathieu Albert, Elise Paradis, and Ayelet Kuper, "Interdisciplinary Promises versus Practices in Medicine: The Decoupled Experiences of Social Sciences and Humanities Scholars," *Social Science & Medicine* 126 (2015): 17-25; Alison Pilnick, "Sociology without Frontiers? On the Pleasures and Perils of Interdisciplinary Research," *Sociological Research Online* 18, no. 3 (2013): 97-104; and Burch et al., "Social Scientist-STEM Collaborations in Agriculture, Food and Beyond."

18. Summer Sullivan and Julie Guthman, "Sensing the Fields: A Report on the Emerging Ag-Tech Initiative at UCSC," UC Santa Cruz, 2022, accessed January 31, 2023, https://escholarship.org/uc/item/6029523m; and Summer Sullivan, "Ag-Tech, Agroecology, and the Politics of Alternative Farming Futures: The Challenges of Bringing Together Diverse Agricultural Epistemologies," *Agriculture and Human Values* 40, no. 3 (2023): 913-28.

## Chapter 7. Big Ideas and Making Silicon Valley–Style Solution-Makers

1. "What's Your Big Idea?," www.bigideascontest.org, accessed February 2022.

2. Giridharadas, *Winners Take All*, 27.

3. "What's Your Big Idea?," emphasis added.

4. Michael Mascarenhas, *New Humanitarianism and the Crisis of Charity: Good Intentions on the Road to Help* (Bloomington: Indiana University Press, 2017).

5. "What's Your Big Idea?," emphasis added.
6. Vinsel, "Design Thinking Movement Is Absurd."
7. Cole, "White-Savior Industrial Complex."
8. Stein et al., "Beyond Colonial Futurities in Climate Education."
9. Cole, "White-Savior Industrial Complex"; Li, *Will to Improve*.
10. Li, *Will to Improve*, 4–5.
11. Stein et al., "Beyond Colonial Futurities in Climate Education."
12. Clare Talwalker, "Fixing Poverty," in *Encountering Poverty: Thinking and Acting in an Unequal World*, ed. Ananya Roy et al. (Oakland: University of California Press, 2016), 133.
13. Benjamin, *Race after Technology*.
14. Cole, "White-Savior Industrial Complex"; and Roy et al., *Encountering Poverty*.
15. Stein et al., "Beyond Colonial Futurities in Climate Education."
16. Stein et al., "Beyond Colonial Futurities in Climate Education."
17. Paolo Freire, *Pedagogy of the Oppressed* (1970; New York: Continuum, 2000), 126.
18. Freire, *Pedagogy of the Oppressed*; and Roy et al., *Encountering Poverty*.
19. INCITE! Women of Color Against Violence, ed., *The Revolution Will Not Be Funded: Beyond the Non-Profit Industrial Complex* (Cambridge, MA: South End Press, 2006).
20. Roy et al., *Encountering Poverty*, 2.
21. Genevieve Negrón-Gonzales, "Teaching Poverty," in *Encountering Poverty: Thinking and Acting in an Unequal World*, ed. Ananya Roy et al. (Oakland: University of California Press, 2016), 158–59.
22. Ananya Roy, "Encountering Poverty," in *Encountering Poverty*, ed. Ananya Roy et al. (Oakland: University of California Press, 2016), 48.
23. Stein et al., "Beyond Colonial Futurities in Climate Education," 993.

## Conclusion

1. Zoe Sayler, "Alphabet's Captain of Moonshots: If No One Laughs, Your Idea Isn't Big Enough," *Grist*, October 23, 2019, https://grist.org/article/alphabets-captain-of-moonshots-if-no-one-laughs-your-idea-isnt-big-enough.
2. Oliver Franklin-Wallis, "Inside X, Google's Top-Secret Moonshot Factory," *Wired*, February 17, 2020, www.wired.co.uk/article/ten-years-of-google-x.

3. New Food Order, "Designing a Climate-Friendly Food Company, with Julia Collins," podcast, December 13, 2022, https://podcasts.apple.com/us/podcast/designing-a-climate-friendly-food-company-with/id1651879872?i=1000590015686.

4. Barbrook and Cameron, "Californian Ideology."

5. Giridharadas, *Winners Take All*.

6. Linsey McGoey, "Strategic Unknowns: Towards a Sociology of Ignorance," *Economy and Society* 41, no. 1 (2012): 1–16.

7. Giridharadas, *Winners Take All*.

8. Mihir Desai, "The Crypto Collapse and the End of the Magical Thinking That Infected Capitalism," *New York Times*, January 16, 2023, www.nytimes.com/2023/01/16/opinion/the-crypto-collapse-magical-thinking-capitalism.html?campaign_id=39.

9. Jenny Odell, *How to Do Nothing: Resisting the Attention Economy* (New York: Melville House, 2020).

10. Besky, *Darjeeling Distinction*.

11. adrienne maree brown, *Emergent Strategy: Shaping Change, Changing Worlds* (Chico, CA: AK Press, 2017); and Marshall Ganz, *Why David Sometimes Wins: Leadership, Organization, and Strategy in the California Farm Worker Movement* (New York: Oxford University Press, 2009).

12. Ruth Wilson Gilmore, foreword to *The Struggle Within: Prisons, Political Prisoners, and Mass Movements in the United States*, ed. Dan Berger (Oakland, CA: PM Press, 2014); and André Gorz, "Strategy for Labor," in *Theories of the Labor Movement* (Detroit, MI: Wayne State University Press, 1987), 102.

13. Harris, *Palo Alto*.

# *Bibliography*

AgFunder. "AgFunder AgriFoodTech Investment Report." 2022. Accessed December 9, 2022. https://agfunder.com/research/2022-agfunder-agrifoodtech-investment-report.

AgThentic. "How Silicon Valley Set Agtech Back a Decade." *AgThentic Blog*, July 6, 2021. https://blog.agthentic.com/how-silicon-valley-set-agtech-back-a-decade-b9d46e0acf0d.

Albert, Mathieu, Elise Paradis, and Ayelet Kuper. "Interdisciplinary Promises versus Practices in Medicine: The Decoupled Experiences of Social Sciences and Humanities Scholars." *Social Science & Medicine* 126 (2015): 17–25.

Andreessen, Marc. "Why Software Is Eating the World." *Wall Street Journal*, August 20, 2011. www.wsj.com/articles/SB10001424053111903480904576512250915629460.

Barbrook, Richard, and Andy Cameron. "Californian Ideology: A Critique of West Coast Cyber-Libertarianism." *Science as Culture* 6, no. 1 (1996): 44–72.

Belasco, Warren J. *Appetite for Change*. New York: Pantheon, 1989.

Bell, Shannon Elizabeth, Alicia Hullinger, and Lilian Brislen. "Manipulated Masculinities: Agribusiness, Deskilling, and the Rise of the Businessman-Farmer in the United States." *Rural Sociology* 80, no. 3 (2015): 285–313.

Benjamin, Ruha. *Race after Technology: Abolitionist Tools for the New Jim Code*. Cambridge, UK: Polity Press, 2019.

Berman, Elizabeth Popp. *Creating the Market University: How Academic Science Became an Economic Engine*. Princeton, NJ: Princeton University Press, 2012.

Besky, Sarah. *The Darjeeling Distinction: Labor and Justice on Fair-Trade Tea Plantations in India*. Berkeley: University of California Press, 2013.

Biltekoff, Charlotte. *Eating Right in America: The Cultural Politics of Food and Health*. Durham, NC: Duke University Press, 2013.

Biltekoff, Charlotte, and Julie Guthman. "Conscious, Complacent, Fearful: Agri-Food Tech's Market-Making Public Imaginaries." *Science as Culture* 32, no. 1 (2023): 58–82.

Blanchette, Alex. *Porkopolis: American Animality, Standardized Life, and the Factory Farm*. Durham, NC: Duke University Press, 2020.

Block, Fred, and Matthew R. Keller. "Where Do Innovations Come From? Transformations in the US Economy, 1970–2006." *Socio-Economic Review* 7, no. 3 (2009): 459–83.

Bolotnikova, Marina. "Five Things to Know about the SCOTUS Challenge to California's Ban on Extreme Farm Animal Confinement." *The Counter*, March 29, 2022. https://thecounter.org/california-proposition-12-animal-welfare-supreme-court.

Borlaug, Norman. "The Nobel Prize: Norman Borlaug Acceptance Speech." December 10, 1970. www.nobelprize.org/prizes/peace/1970/borlaug/acceptance-speech.

Boyd, William, and Michael J. Watts. "Agro-Industrial Just-in-Time: The Chicken Industry and Postwar American Capitalism." In *Globalising Food: Agrarian Questions and Global Restructuring*, edited by David Goodman and Michael J. Watts, 192–225. London: Routledge, 1997.

Breakthrough Institute. "An Ecomodernist Manifesto." 2015. Accessed December 23, 2019. https://ecomodernistmanifesto.squarespace.com.

Broad, Garrett M. "Improving the Agri-Food Biotechnology Conversation: Bridging Science Communication with Science and Technology Studies." *Agriculture and Human Values* 40, no. 3 (2023): 929–38.

Bronson, Kelly. "Excluding 'Anti-Biotech' Activists from Canadian Agri-Food Policy Making: Ethical Implications of the Deficit Model of Science Communication." In *Ethics and Practice in Science Communication*, edited by Susanna Priest, Jean Goodwin, and Michael F. Dahlstrom, 235–52. Chicago: University of Chicago Press, 2018.

———. *The Immaculate Conception of Data: Agribusiness, Activists, and Their Shared Politics of the Future*. Montreal: McGill-Queen's University Press, 2022.

———. "Looking through a Responsible Innovation Lens at Uneven Engagements with Digital Farming." *NJAS: Wageningen Journal of Life Sciences* 90–91, no. 1 (2019): 1–6.

Bronson, Kelly, and Irena Knezevic. "Big Data in Food and Agriculture." *Big Data & Society* 3, no. 1 (2016). https://doi.org/10.1177/2053951716648174.

brown, adrienne maree. *Emergent Strategy: Shaping Change, Changing Worlds.* Chico, CA: AK Press, 2017.

Brown, Tim. "Design Thinking." *Harvard Business Review* (June 2008). https://hbr.org/2008/06/design-thinking.

Brown, Tim, and Barry Katz. "Change by Design." *Journal of Product Innovation Management* 28, no. 3 (2011): 381–83.

Bryant, Raymond L., and Michael K. Goodman. "Consuming Narratives: The Political Ecology of 'Alternative' Consumption." *Transactions of the Institute of British Geographers* 29, no. 3 (2004): 344–66.

Buchanan, Richard. "Wicked Problems in Design Thinking." *Design Issues* 8, no. 2 (1992): 5–21.

Burch, Karly, Julie Guthman, Mascha Gugganig, Kelly Bronson, Matt Comi, Katharine Legun, Charlotte Biltekoff, et al. "Social Science–STEM Collaborations in Agriculture, Food and Beyond: An STSFAN Manifesto." *Agriculture and Human Values* 40, no. 3 (2023): 939–49.

Burwood-Taylor, Louisa. "Eighteen94 Capital Leads Funding in Kuli Kuli." *AgFunder News*, January 17, 2017. https://agfundernews.com/breaking-kelloggs-vc-1894-capital-makes-first-investment-agfunder-alum-kuli-kuli.html.

Busch, Lawrence. *Knowledge for Sale: The Neoliberal Takeover of Higher Education.* Cambridge, MA: MIT Press, 2017.

Busch, Lawrence, William B. Lacy, Jeffrey Burkhardt, and Laura R. Lacy. *Plants, Power, and Profit.* Cambridge, MA: Blackwell, 1991.

Buttel, Frederick H. "The Recombinant BGH Controversy in the United States: Toward a New Consumption Politics of Food?" *Agriculture and Human Values* 17, no. 1 (2000): 5–20.

Buttel, Frederick H., and Lawrence Busch. "The Public Agricultural Research System at the Crossroads." *Agricultural History* 62, no. 2 (1988): 303–24.

Carpenter, Kenneth J. "The History of Enthusiasm for Protein." *Journal of Nutrition* 116, no. 7 (1986): 1364–70.

Christensen, Clayton M., Michael E. Raynor, and Rory McDonald. "What Is Disruptive Innovation?" *Harvard Business Review* (December 2015). https://hbr.org/2015/12/what-is-disruptive-innovation.

Clark, Cathy, Jed Emerson, and Ben Thornley. *Collaborative Capitalism and the Rise of Impact Investing*. San Francisco: Jossey-Bass, 2014.

Cochrane, Willard W. *The Development of American Agriculture*. Minneapolis: University of Minnesota Press, 1993.

Cohen, Noam. "M.I.T. Media Lab, Already Rattled by the Epstein Scandal, Has a New Worry." *New York Times*, September 22, 2019. www.nytimes.com/2019/09/22/business/media/mit-media-lab-food-computer.html.

Cohen, Susan, and Yael V. Hochberg. "Accelerating Startups: The Seed Accelerator Phenomenon." *SSRN Electronic Journal* (March 30, 2014). https://doi.org/10.2139/ssrn.2418000.

Cole, Teju. "The White-Savior Industrial Complex." *The Atlantic*, March 21, 2012. www.theatlantic.com/international/archive/2012/03/the-white-savior-industrial-complex/254843.

Cotter, Joseph. *Troubled Harvest: Agronomy and Revolution in Mexico, 1880–2002*. Westport, CT: Greenwood Publishing Group, 2003.

Creswell, Julie. "Beyond Meat Is Struggling, and the Plant-Based Meat Industry Worries." *New York Times*, November 21, 2022. www.nytimes.com/2022/11/21/business/beyond-meat-industry.html.

Cullather, Nick. *The Hungry World: America's Cold War Battle against Poverty in Asia*. Cambridge, MA: Harvard University Press, 2010.

Desai, Mihir. "The Crypto Collapse and the End of the Magical Thinking That Infected Capitalism." *New York Times*, January 16, 2023. www.nytimes.com/2023/01/16/opinion/the-crypto-collapse-magical-thinking-capitalism.html?campaign_id=39.

Devereux, Stephen. *The New Famines: Why Famines Persist in an Era of Globalization*. London: Routledge, 2006.

———. "Sen's Entitlement Approach: Critiques and Counter-Critiques." *Oxford Development Studies* 29, no. 3 (2001): 245–63.

De Waal, Alex. *Mass Starvation: The History and Future of Famine*. Hoboken, NJ: John Wiley & Sons, 2017.

Duncan, Emily, Alesandros Glaros, Dennis Z. Ross, and Eric Nost. "New but for Whom? Discourses of Innovation in Precision Agriculture." *Agriculture and Human Values* 38 (2021): 1181–99.

Easterbrook, Gregg. "Forgotten Benefactor of Humanity." *The Atlantic*, January 1997. www.theatlantic.com/magazine/archive/1997/01/forgotten-benefactor-of-humanity/306101.

Eilperin, Juliet. "Why the Clean Tech Boom Went Bust." *Wired*, January 20, 2012. www.wired.com/2012/01/ff_solyndra.

Fairbairn, Madeleine, and Zenia Kish. "'A Poverty of Data'? Exporting the Digital Revolution to Farmers in the Global South." In *The Nature of Data: Infrastructures, Environments, Politics*, edited by Jenny Goldstein and Eric Nost, 211–29. Lincoln: University of Nebraska Press, 2022.

Fairbairn, Madeleine, and Emily Reisman. "The Incumbent Advantage: Corporate Power in Agri-Food Tech." *Journal of Peasant Studies* (2024). https://doi.org/10.1080/03066150.2024.2310146.

Ferenstein, Gregory. "Silicon Valley's New Politics of Optimism, Radical Idealism and Bizarre Loyalties." *The Guardian*, November 10, 2015. www.theguardian.com/us-news/2015/nov/10/silicon-valley-politics-tech-industry.

Ferguson, James. *The Anti-Politics Machine: Development, Depoliticization, and Bureaucratic Power in Lesotho*. Minneapolis: University of Minnesota Press, 1994.

Fielke, Simon, Kelly Bronson, Michael Carolan, Callum Eastwood, Vaughan Higgins, Emma Jakku, Laurens Klerkx, et al. "A Call to Expand Disciplinary Boundaries So That Social Scientific Imagination and Practice Are Central to Quests for 'Responsible' Digital Agri-Food Innovation." *Sociologia Ruralis* 62, no. 2 (2022): 151–61.

Fine, Ben. "Towards a Political Economy of Food." *Review of International Political Economy* 1, no. 3 (1994): 519–45.

FitzSimmons, Margaret, and David Goodman. "Incorporating Nature: Environmental Narratives and the Reproduction of Food." In *Remaking Reality: Nature at the Millenium*, edited by Bruce Braun and Noel Castree, 194–220. London: Routledge, 1998.

Fox, Nick J. "Green Capitalism, Climate Change and the Technological Fix: A More-Than-Human Assessment." *Sociological Review* 71, no. 5 (2022). https://doi.org/10.1177/00380261221121232.

Franklin-Wallis, Oliver. "Inside X, Google's Top-Secret Moonshot Factory." *Wired*, February 17, 2020. www.wired.co.uk/article/ten-years-of-google-x.

Fraser, Alistair. "Land Grab/Data Grab: Precision Agriculture and Its New Horizons." *Journal of Peasant Studies* 46, no. 5 (2019): 893–912.

Freidberg, Susanne. "Big Food and Little Data: The Slow Harvest of Corporate Food Supply Chain Sustainability Initiatives." *Annals of the American Association of Geographers* 107, no. 6 (2017): 1389–406.

Freire, Paolo. *Pedagogy of the Oppressed*. 1970; New York: Continuum, 2000.

Friedland, William. "The End of Rural Society and the Future of Rural Sociology." Paper prepared for the Annual Meeting of the Rural Sociological Society, Guelph, Ontario, 1981. https://files.eric.ed.gov/fulltext/ED211319.pdf.

Friedmann, Harriet. "Distance and Durability: Shaky Foundations of the World Food Economy." In *The Global Restructuring of Agro-Food Systems*, edited by Philip McMichael, 258–76. Ithaca, NY: Cornell University Press, 1994.

Ganz, Marshall. *Why David Sometimes Wins: Leadership, Organization, and Strategy in the California Farm Worker Movement*. New York: Oxford University Press, 2009.

Geiger, Susi. "Silicon Valley, Disruption, and the End of Uncertainty." *Journal of Cultural Economy* 13, no. 2 (2020): 169–84.

Gianella, Eric. "Morality and the Idea of Progress in Silicon Valley." *Berkeley Journal of Sociology* (January 14, 2015). http://berkeleyjournal.org/2015/01/morality-and-the-idea-of-progress-in-silicon-valley.

Gilmore, Ruth Wilson. Foreword to *The Struggle Within: Prisons, Political Prisoners, and Mass Movements in the United States*, by Dan Berger, vii–viii. Oakland, CA: PM Press, 2014.

Giridharadas, Anand. *Winners Take All: The Elite Charade of Changing the World*. New York: Vintage Books, 2018.

Glenna, Leland L., William B. Lacy, Rick Welsh, and Dina Biscotti. "University Administrators, Agricultural Biotechnology, and Academic Capitalism: Defining the Public Good to Promote University-Industry Relationships." *Sociological Quarterly* 48, no. 1 (2007): 141–63.

Goldstein, Harry. "MIT Media Lab's Food Computer Project Permanently Shut Down." *IEEE Spectrum*, May 17, 2020. https://spectrum.ieee.org/mit-media-lab-food-computer-project-shut-down.

Goldstein, Jesse. *Planetary Improvement: Cleantech Entrepreneurship and the Contradictions of Green Capitalism*. Cambridge, MA: MIT Press, 2018.

Goodman, David, and Michael Redclift. "Constructing a Political Economy of Food." *Review of International Political Economy* 1, no. 3 (1994): 547–52.

Goodman, David, Bernardo Sorj, and John Wilkinson. *From Farming to Biotechnology*. Oxford, UK: Basil Blackwell, 1987.

Goody, Jack. "Industrial Food: Towards the Development of a World Cuisine." In *Food and Culture: A Reader*, edited by Carole Counihan and Penny Van Esterik, 86–104. New York: Routledge, 2012.

Gunn, Andrew, and Michael Mintrom. "Evaluating the Non-Academic Impact of Academic Research: Design Considerations." *Journal of Higher Education Policy and Management* 39, no. 1 (2017): 20–30.

Guthman, Julie. "Binging and Purging: Agrofood Capitalism and the Body as Socioecological Fix." *Environment and Planning A* 47, no. 12 (2015): 2522–36.

———. *Wilted: Pathogens, Chemicals, and the Fragile Future of the Strawberry Industry*. Oakland: University of California Press, 2019.

Guthman, Julie, and Charlotte Biltekoff. "Agri-Food Tech's Building Block: Narrating Protein, Agnostic of Source, in the Face of Crisis." *BioSocieties* 18 (2023): 656–78.

———. "Magical Disruption? Alternative Protein and the Promise of De-Materialization." *Environment and Planning E: Nature and Space* 4, no. 4 (2021): 1583–600.

Guthman, Julie, and Michaelanne Butler. "Fixing Food with a Limited Menu: On (Digital) Solutionism in the Agri-Food Tech Sector." *Agriculture and Human Values* 40, no. 3 (2023): 835–48.

Hajer, Maarten A. *The Politics of Environmental Discourse: Ecological Modernization and the Policy Process*. New York: Oxford University Press, 1995.

Hall, Stuart. "The West and the Rest: Discourse and Power." In *Formations of Modernity*, edited by Stuart Hall and Bram Gieben, 275–320. Cambridge, UK: Polity Press, 1992.

Harris, Malcom. *Palo Alto: A History of California, Capitalism, and the World*. New York: Little, Brown & Co., 2023.

Harrison, Jill Lindsey. *From the Inside Out: The Fight for Environmental Justice within Government Agencies*. Cambridge, MA: MIT Press, 2019.

Harvey, David. *A Brief History of Neoliberalism*. New York: Oxford University Press, 2005.

Hedberg, Russell. "Bad Animals, Techno-Fixes, and the Environmental Narratives of Alternative Protein." *Frontiers in Sustainable Food Systems* 7 (2023). https://doi.org/10.3389/fsufs.2023.1160458.

Henke, Christopher R. *Cultivating Science, Harvesting Power*. Cambridge, MA: MIT Press, 2008.

Hetherington, Kregg. *The Government of Beans: Regulating Life in the Age of Monocrops*. Durham, NC: Duke University Press, 2020.

Hill, Steven. "Assessing (for) Impact: Future Assessment of the Societal Impact of Research." *Palgrave Communications* 2, no. 1 (2016): 1–7.

Hinchliffe, Steve, Nick Bingham, John Allen, and Simon Carter. *Pathological Lives: Disease, Space and Biopolitics*. Chichester, UK: John Wiley & Sons, 2016.

Hogarth, Stuart. "Valley of the Unicorns: Consumer Genomics, Venture Capital and Digital Disruption." *New Genetics and Society* 36, no. 3 (2017): 250–72.

Howard, Philip H. *Concentration and Power in the Food System: Who Controls What We Eat?* London: Bloomsbury Publishing, 2016.

Howard, Philip H., Francesco Ajena, Marina Yamaoka, and Amber Clarke. "'Protein' Industry Convergence and Its Implications for Resilient and Equitable Food Systems." *Frontiers in Sustainable Food Systems* 5 (2021). https://doi.org/10.3389/fsufs.2021.684181.

Huesemann, Michael, and Joyce Huesemann. *Techno-Fix: Why Technology Won't Save Us or the Environment*. Gabriola Island, BC: New Society Publishers, 2011.

INCITE! Women of Color Against Violence, ed. *The Revolution Will Not Be Funded: Beyond the Non-Profit Industrial Complex*. Cambridge, MA: South End Press, 2006.

Isaac, Marney E., S. Ryan Isakson, Bryan Dale, Charles Z. Levkoe, Sarah K. Hargreaves, V. Ernesto Méndez, Hannah Wittman, et al. "Agroecology in Canada: Towards an Integration of Agroecological Practice, Movement, and Science." *Sustainability* 10, no. 9 (2018): https://doi.org/10.3390/su10093299.

Jacobs, Jerry A., and Scott Frickel. "Interdisciplinarity: A Critical Assessment." *Annual Review of Sociology* 35, no. 1 (2009): 43–65.

Jasanoff, Sheila. *States of Knowledge: The Co-Production of Science and Social Order*. London: Routledge, 2004.

Jervis, Francis. "Eating the World: Iterative Capital after Silicon Valley." PhD dissertation, New York University, 2020.

Johnson, Marilynn. *The Second Gold Rush: Oakland and the East Bay in World War II*. Berkeley: University of California, 1994.

Johnston, Sean F. "Alvin Weinberg and the Promotion of the Technological Fix." *Technology and Culture* 59, no. 3 (2018): 620–51.

Jönsson, Erik. "Benevolent Technotopias and Hitherto Unimaginable Meats: Tracing the Promises of In Vitro Meat." *Social Studies of Science* 46, no. 5 (2016): 725–48.

Jönsson, Erik, Tobias Linné, and Ally McCrow-Young. "Many Meats and Many Milks? The Ontological Politics of a Proposed Post-Animal Revolution." *Science as Culture* 28, no. 1 (2019): 70–97.

Kautsky, Karl. *The Agrarian Question*. 1899; London: Zwan Press, 1988.

Kearney, Michael. "Cleantech's Comeback." *The Engine*, November 9, 2020. https://medium.com/tough-tech/cleantechs-comeback-38ec13507452.

Keerie, Maia. "Record $3.1 Billion Invested in Alt Proteins in 2020 Signals Growing Market Momentum for Sustainable Proteins." Good Food Institute, March 18, 2021. https://gfi.org/blog/2020-state-of-the-industry-highlights.

Kilman, Scott, and Roger Thurow. "Father of 'Green Revolution' Dies." *Wall Street Journal*, September 13, 2009. www.wsj.com/articles/SB125281643150406425.

Kimura, Aya Hirata. *Hidden Hunger: Gender and the Politics of Smarter Foods*. Ithaca, NY: Cornell University Press, 2013.

Klein, Naomi. *This Changes Everything: Capitalism vs. the Climate*. New York: Simon and Schuster, 2015.

Klerkx, Laurens, and David Rose. "Dealing with the Game-Changing Technologies of Agriculture 4.0: How Do We Manage Diversity and Responsibility in Food System Transition Pathways?" *Global Food Security* 24 (2020). https://doi.org/10.1016/j.gfs.2019.100347.

Kloppenburg, Jack. *First the Seed: The Political Economy of Plant Biotechnology*. Madison: University of Wisconsin Press, 2005.

Kremen, Claire, Alastair Iles, and Christopher Bacon. "Diversified Farming Systems: An Agroecological, Systems-Based Alternative to Modern Industrial Agriculture." *Ecology and Society* 17, no. 4 (2012). http://dx.doi.org/10.5751/ES-05103-170444.

Kuli Kuli. "About Us." Accessed January 2, 2019. www.kulikulifoods.com/about.

Lam, A. C. Y., A. Can Karaca, R. T. Tyler, and M. T. Nickerson. "Pea Protein Isolates: Structure, Extraction, and Functionality." *Food Reviews International* 34, no. 2 (2018): 126–47.

Landecker, Hannah. "Antibiotic Resistance and the Biology of History." *Body & Society* 22, no. 4 (2015): 1–34.

Lappé, Frances Moore. *Diet for a Small Planet*. New York: Ballantine Books, 1971.

Laskow, Sarah. "A Distant Voyage, a Powerful Plant, and a Crowd-Backed Quest to Crack the Snack Market." *Fast Company*, April 24, 2014. www.fastcompany.com/3029485/the-amazing-plant-powering-a-quest-to-crack-the-crowded-snack-market.

Leach, Melissa. *Rainforest Relations*. Washington, DC: Smithsonian Institution Press, 1994.

Lepore, Jill. "The Disruption Machine." *The New Yorker*, June 16, 2014, 30–36.

Levenstein, Harvey A. *Paradox of Plenty: A Social History of Eating in Modern America*. New York: Oxford University Press, 1993.

Levidow, Les, Michel Pimbert, and Gaetan Vanloqueren. "Agroecological Research: Conforming—or ransforming the Dominant Agro-Food Regime?" *Agroecology and Sustainable Food Systems* 38, no. 10 (2014): 1127–55.

Li, Tania Murray. *The Will to Improve: Governmentality, Development, and the Practice of Politics*. Durham, NC: Duke University Press, 2007.

Liboiron, Max. "Modern Waste Is an Economic Strategy." *Lo Squaderno: Explorations in Space and Society* (special edition on Garbage and Wastes) 29 (2013): 9–12.

Lien, Marianne Elisabeth. *Becoming Salmon: Aquaculture and the Domestication of a Fish*. Oakland: University of California Press, 2015.

Lioutas, Evagelos D., and Chrysanthi Charatsari. "Big Data in Agriculture: Does the New Oil Lead to Sustainability?" *Geoforum* 109 (2020): 1–3.

Lorimer, Jamie. "Probiotic Environmentalities: Rewilding with Wolves and Worms." *Theory, Culture & Society* 34, no. 4 (2017): 27–48.

Lowe, Philip, Jonathan Murdoch, Terry Marsden, Richard Munton, and Andrew Flynn. "Regulating the New Rural Spaces: The Uneven Development of Land." *Journal of Rural Studies* 9, no. 3 (1993): 205–22.

Lynch, John, and Raymond Pierrehumbert. "Climate Impacts of Cultured Meat and Beef Cattle." *Frontiers in Sustainable Food Systems* 3 (2019). https://doi.org/10.3389/fsufs.2019.00005.

MacKenzie, Donald. "Is Economics Performative? Option Theory and the Construction of Derivatives Markets." *Journal of the History of Economic Thought* 28, no. 1 (2006): 29–55.

Malthus, Thomas Robert. *An Essay on the Principle of Population.* Edited by Geoffrey Gilbert. Oxford, UK: Oxford University Press, 1999.

Mann, Charles. "Can Planet Earth Feed 10 Billion People?" *The Atlantic,* March 2018. www.theatlantic.com/magazine/archive/2018/03/charles-mann-can-planet-earth-feed-10-billion-people/550928.

Mann, Geoff. *Disassembly Required: A Field Guide to Actually Existing Capitalism.* Oakland, CA: AK Press, 2013.

Mann, Susan A. *Agrarian Capitalism in Theory and Practice.* Chapel Hill: University of North Carolina Press, 1990.

Mascarenhas, Michael. *New Humanitarianism and the Crisis of Charity: Good Intentions on the Road to Help.* Bloomington: Indiana University Press, 2017.

Mattick, Carolyn S., Amy E. Landis, Braden R. Allenby, and Nicholas J. Genovese. "Anticipatory Life Cycle Analysis of In Vitro Biomass Cultivation for Cultured Meat Production in the United States." *Environmental Science & Technology* 49, no. 19 (2015): 11941–49.

Mazzucato, Mariana. *The Entrepreneurial State: Debunking Public vs. Private Sector Myths.* New York: Public Affairs, 2015.

McFettridge, Scott. "The Weird State of Organic Farming in the U.S.: Consumers Love It, but Fewer and Fewer Farmers Are Converting." *Fortune,* September 22, 2022. https://fortune.com/2022/09/22/organic-farming-popular-but-not-with-farmers-converting.

McGoey, Linsey. "Strategic Unknowns: Towards a Sociology of Ignorance." *Economy and Society* 41, no. 1 (2012): 1–16.

McKay, Tom. "MIT Built a Theranos for Plants." *Gizmodo,* September 8, 2019. https://gizmodo.com/mit-built-a-theranos-for-plants-1837968240.

Memmi, Albert. *The Colonizer and the Colonized.* Boston: Orion Press, 1967.

Mirowski, Philip. *Science-Mart: Privatizing American Science.* Cambridge, MA: Harvard University Press, 2011.

Mitchell, Katharyne, and Matthew Sparke. "The New Washington Consensus: Millennial Philanthropy and the Making of Global Market Subjects." *Antipode* 48, no. 3 (2016): 724–49.

Mitchell, Timothy. *Rule of Experts: Egypt, Techno-Politics, Modernity*. Berkeley: University of California Press, 2002.

MIT Media Lab. "Launching the OpenAG Initiative at the MIT Media Lab." *Medium.com*, October 16, 2015. https://medium.com/mit-media-lab/launching-the-openag-initiative-at-the-mit-media-lab-7547f0d994a6.

Morozov, Evgeny. *To Save Everything, Click Here: Technology, Solutionism, and the Urge to Fix Problems That Don't Exist*. London: Penguin, 2013.

Mouat, Michael J., and Russell Prince. "Cultured Meat and Cowless Milk: On Making Markets for Animal-Free Food." *Journal of Cultural Economy* 11, no. 4 (2018): 315–29.

Mulvaney, Dustin. *Solar Power: Innovation, Sustainability, and Environmental Justice*. Oakland: University of California Press, 2019.

Naess, A. "The Shallow and the Deep, Long-Range Ecology Movement. A Summary." *Inquiry* 16 (1973): 95–100.

Negrón-Gonzales, Genevieve. "Teaching Poverty." In *Encountering Poverty: Thinking and Acting in an Unequal World*, edited by Ananya Roy, Genevieve Negrón-Gonzales, Kweku Opoku-Agyemang, and Clare Talwalker, 149–76. Oakland: University of California Press, 2016.

New Food Order. "Designing a Climate-Friendly Food Company, with Julia Collins." Podcast. December 13, 2022. https://podcasts.apple.com/us/podcast/designing-a-climate-friendly-food-company-with/id1651879872?i=1000590015686.

Niazi, Tarique. "Rural Poverty and the Green Revolution: The Lessons from Pakistan." *Journal of Peasant Studies* 31, no. 2 (2004): 242–60.

Nicholas, Tom. *VC: An American History*. Cambridge, MA: Harvard University Press, 2019.

Norgaard, Kari. *Living in Denial: Climate Change, Emotions, and Everyday Life*. Cambridge, MA: MIT Press, 2011.

Odell, Jenny. *How to Do Nothing: Resisting the Attention Economy*. New York: Melville House, 2020.

O'Mara, Margaret. *The Code: Silicon Valley and the Remaking of America*. New York: Penguin, 2020.

O'Rourke, Anastasia Rose. "The Emergence of Cleantech." PhD dissertation, Yale University, 2009.

Orr, C. "Academic Activism." In *The Women's Movement Today: An Encyclopedia of Third Wave Feminism*, edited by Leslie L. Heywood, 1–4. Westport, CT: Greenwood, 2006.

Paquet, Gilles. *The New Geo-Governance: A Baroque Approach*. Ottawa: University of Ottawa Press, 2005.

Paratore, Michelle. "Kuli Kuli: The Next Superfood, and a Way to Support Women in West Africa." *Edible Startups*, December 4, 2013. https://ediblestartups.com/2013/12/04/kuli-kuli-the-next-superfood-and-a-way-to-support-women-in-west-africa.

Patel, Raj. "The Long Green Revolution." *Journal of Peasant Studies* 40, no. 1 (2013): 1–63.

Perkins, John H. "The Rockefeller Foundation and the Green Revolution, 1941–1956." *Agriculture and Human Values* 7, no. 3 (1990): 6–18.

Pilnick, Alison. "Sociology without Frontiers? On the Pleasures and Perils of Interdisciplinary Research." *Sociological Research Online* 18, no. 3 (2013): 97–104.

Pingali, Prabhu L. "Green Revolution: Impacts, Limits, and the Path Ahead." *Proceedings of the National Academy of Sciences* 109, no. 31 (2012).

Poppendieck, Janet. *Breadlines Knee Deep in Wheat: Food Assistance in the Great Depression*. New Brunswick, NJ: Rutgers University Press, 1986.

Prause, Louisa, Sarah Hackfort, and Margit Lindgren. "Digitalization and the Third Food Regime." *Agriculture and Human Values* 38, no. 3 (2021): 641–55.

Reiley, Laura. "Alt-Meat Fever Has Cooled. Here's Why." *Washington Post*, November 12, 2022. www.washingtonpost.com/business/2022/11/12/plant-based-meat-market.

Reisman, Emily. "Sanitizing Agri-Food Tech: Covid-19 and the Politics of Expectation." *Journal of Peasant Studies* 48, no. 5 (2021): 910–33.

Reisman, Jane, Veronica Olazabal, and Shawna Hoffman. "Putting the 'Impact' in Impact Investing: The Rising Demand for Data and Evidence of Social Outcomes." *American Journal of Evaluation* 39, no. 3 (2018): 389–95.

Rhodes, Carl, Christopher Wright, and Alison Pullen. "Changing the World? The Politics of Activism and Impact in the Neoliberal University." *Organization* 25, no. 1 (2018): 139–47.

Robinson, Cedric. *Black Marxism: The Making of the Black Radical Tradition*. 3rd edition. Chapel Hill: University of North Carolina, 2021.

Rojek, Chris. "'Big Citizen' Celanthropy and Its Discontents." *International Journal of Cultural Studies* 17, no. 2 (2014): 127–41.

Ross, Eric B. *The Malthus Factor: Poverty, Politics, and Population in Capitalist Development.* London: Zed Books, 1998.

Roy, Ananya. "Encountering Poverty." In *Encountering Poverty: Thinking and Acting in an Unequal World*, edited by Ananya Roy, Genevieve Negrón-Gonzales, Kweku Opoku-Agyemang, and Clare Talwalker, 21–50. Oakland: University of California Press, 2016.

———. "Governing Poverty." In *Encountering Poverty: Thinking and Acting in an Unequal World*, edited by Ananya Roy, Genevieve Negrón-Gonzales, Kweku Opoku-Agyemang, and Clare Talwalker, 51–90. Oakland: University of California Press, 2016.

———. *Poverty Capital: Microfinance and the Making of Development.* New York: Routledge, 2010.

Roy, Ananya, Genevieve Negrón-Gonzales, Kweku Opoku-Agyemang, and Clare Talwalker, eds. *Encountering Poverty: Thinking and Acting in an Unequal World.* Oakland: University of California Press, 2016.

Rudy, Alan P., Dawn Coppin, Jason Konefal, Bradley T. Shaw, Toby Van Eyck, Craig Harris, and Lawrence Busch. *Universities in the Age of Corporate Science: The UC Berkeley-Novartis Controversy.* Philadelphia: Temple University Press, 2007.

Saxenian, AnnaLee. *Regional Advantage: Culture and Competition in Silicon Valley and Route 128.* Cambridge, MA: Harvard University Press, 1996.

Sayler, Zoe. "Alphabet's Captain of Moonshots: If No One Laughs, Your Idea Isn't Big Enough." *Grist*, October 23, 2019. https://grist.org/article/alphabets-captain-of-moonshots-if-no-one-laughs-your-idea-isnt-big-enough.

Schroeder, Richard. "Shady Practice—Gender and the Political Ecology of Resource Stabilization in Gambian Garden Orchards." *Economic Geography* 69, no. 4 (1993): 349–65.

Schurman, Rachel, and William Munro. *Fighting for the Future of Food: Activists versus Agribusiness in the Struggle over Biotechnology.* Minneapolis: University of Minnesota Press, 2010.

Scrinis, Gyorgy, and Carlos Augusto Monteiro. "Ultra Processed Foods and the Limits of Product Reformulation." *Public Health Nutrition* 21, no. 1 (2018): 247–52.

Segal, Howard P. "Practical Utopias: America as Techno-Fix Nation." *Utopian Studies* 28, no. 2 (2017): 231–46.

Seiffert, Don. "MIT Silent over Project Shutdown, Policies." *Boston Business Journal*, October 11, 2022. www.bizjournals.com/boston/news/2022/08/11/mit-silent-over-project-shutdown-policies.html.

Seitz, Tim. "The 'Design Thinking' Delusion." *Jacobin*, October 16, 2018. https://jacobin.com/2018/10/design-thinking-innovation-consulting-politics.

Sen, Amartya. *Poverty and Famines: An Essay on Entitlement and Deprivation*. New York: Oxford University Press, 1982.

Severson, Kim. "The New Secret Chicken Recipe? Animal Cells." *New York Times*, February 15, 2022. www.nytimes.com/2022/02/15/dining/cell-cultured-meat.html.

Sexton, Alexandra E., Tara Garnett, and Jamie Lorimer. "Framing the Future of Food: The Contested Promises of Alternative Proteins." *Environment and Planning E: Nature and Space* 2, no. 1 (2019): 47–72.

Shah, Niyati. "Open Agriculture Initiative: Is Digital Farming the Future of Food?" *Foodtank*, May 2016. https://foodtank.com/news/2016/05/open-agriculture-initiative-digital-farming.

Smith, Robert H. "Livestock Production: The Unsustainable Environmental and Economic Effects of an Industry Out of Control." *Buffalo International Law Journal* 4, no. 1 (1996): 45–129.

Stein, Sharon, Vanessa Andreotti, Cash Ahenakew, Rene Suša, Will Valley, Ninawa Huni Kui, Mateus Tremembé, et al. "Beyond Colonial Futurities in Climate Education." *Teaching in Higher Education* 28, no. 5 (2023): 987–1004.

Steinfeld, Henning, Pierre Gerber, Tom Wassenaar, Vincent Castel, Mauricio Rosales, and Cees de Haan. "Livestock's Long Shadow: Environmental Issues and Options." Food and Agriculture Organization, 2006. Accessed March 23, 2023. www.fao.org/3/a0701e/a0701e00.htm.

Stengers, Isabelle. *Another Science Is Possible: A Manifesto for Slow Science*. Cambridge, UK: Polity Press, 2018.

Stilgoe, Jack, Richard Owen, and Phil Macnaghten. "Developing a Framework for Responsible Innovation." In *The Ethics of Nanotechnology, Geoengineering and Clean Energy*, edited by Andrew Maynard and Jack Stilgoe, 347–59. London: Routledge, 2020.

Striffler, Steve. *Chicken: The Dangerous Transformation of America's Favorite Food*. New Haven, CT: Yale University Press, 2005.

Sullivan, Summer. "Ag-Tech, Agroecology, and the Politics of Alternative Farming Futures: The Challenges of Bringing Together Diverse Agricultural Epistemologies." *Agriculture and Human Values* 40, no. 3 (2023): 913–28.

Sullivan, Summer, and Julie Guthman. "Sensing the Fields: A Report on the Emerging Ag-Tech Initiative at UCSC." UC Santa Cruz, 2022. https://escholarship.org/uc/item/6029523m.

Talwalker, Clare. "Fixing Poverty." In *Encountering Poverty: Thinking and Acting in an Unequal World*, edited by Ananya Roy, Genevieve Negrón-Gonzales, Kweku Opoku-Agyemang, and Clare Talwalker, 121–48. Oakland: University of California Press, 2016.

Target. "Target Launches Collaboration with MIT's Media Lab and IDEO to Explore the Future of Food." News release, October 19, 2015. https://corporate.target.com/article/2015/10/mit-media-lab-collaboration.

Teague, Richard, and Urs Kreuter. "Managing Grazing to Restore Soil Health, Ecosystem Function, and Ecosystem Services." *Frontiers in Sustainable Food Systems* 4 (2020). https://doi.org/10.3389/fsufs.2020.534187.

Tsai, Luke. "Kuli Kuli: Oakland Startup Touts West African 'Superfood.'" *East Bay Express*, December 10, 2013. www.eastbayexpress.com/WhatTheFork/archives/2013/12/10/kuli-kuli-oakland-startup-touts-west-african-superfood.

Tsing, Anna. "Inside the Economy of Appearances." *Public Culture* 12, no. 1 (2000): 115–44.

Tsotsis, Alexia. "Monsanto Buys Weather Big Data Company Climate Corporation for around $1.1b." *TechCrunch*, October 2, 2013. https://techcrunch.com/2013/10/02/monsanto-acquires-weather-big-data-company-climate-corporation-for-930m/?guccounter=1.

Turner, Fred. *From Counterculture to Cyberculture: Stewart Brand, the Whole Earth Network, and the Rise of Digital Utopianism*. Chicago: Chicago University Press, 2006.

Vinsel, Lee. "The Design Thinking Movement Is Absurd." *STS-News.medium.com*, November 26, 2018. https://sts-news.medium.com/the-design-thinking-movement-is-absurd-83df815b92ea.

Visser, Oane, Sarah Ruth Sippel, and Louis Thiemann. "Imprecision Farming? Examining the (in)Accuracy and Risks of Digital Agriculture." *Journal of Rural Studies* 86 (2021): 623–32.

Wagner, David. *What's Love Got to Do with It? A Critical Look at American Charity.* New York: New Press, 2000.
Weis, Tony. *The Ecological Hoofprint: The Global Burden of Industrial Livestock.* London: Zed Books, 2013.
Wik, M., P. Pingali, and S. Broca. "Background Paper for the World Development Report 2008: Global Agricultural Performance: Past Trends and Future Prospects." World Bank, Washington, DC, 2008.
Winders, Bill. *The Politics of Food Supply: U.S. Agricultural Policy in the World Economy.* New Haven, CT: Yale University Press, 2009.
Wolf, Eric. *Europe and the People without History.* Berkeley: University of California Press, 1982.
Wolf, Steven A., and Frederick H. Buttel. "The Political Economy of Precision Farming." *American Journal of Agricultural Economics* 78, no. 5 (1996): 1269–74.
Wolf, Steven A., and Spencer D. Wood. "Precision Farming: Environmental Legitimation, Commodification of Information, and Industrial Coordination." *Rural Sociology* 62, no. 2 (1997): 180–206.
Wurgaft, Benjamin. *Meat Planet: Artificial Flesh and the Future of Food.* Oakland: University of California Press, 2019.
Wyse, Rob. "Impact Investing Defined." *HuffPost*, August 30, 2011. www.huffpost.com/entry/impact-investing-defined_b_941916.
Yong, Ed. "America Is Trapped in a Pandemic Spiral." *The Atlantic*, September 19, 2020. www.theatlantic.com/health/archive/2020/09/pandemic-intuition-nightmare-spiral-winter/616204.
Yuan, LinYee. "MIT OpenAg Releases the Personal Food Computer 3.0, a Stem-Friendly Collaboration with Educators." *Mold*, October 24, 2018. https://thisismold.com/space/farm-systems/mit-openag-releases-the-personal-food-computer-3-0-a-stem-friendly-collaboration-with-educators#.W_MI6dMvzBL.

# Index

academia. *See* universities and colleges
academic capitalism, 147
accelerators, 46-47
activism. *See* organizing and activism
Africa: advocacy for Green Revolution-type projects in, 60; famines in, 63, 64
AgFunder investment analyses, 80, 90-91, 92-93*table*
agrarian reform. *See* land reform
agribusiness, 88-89; alternative proteins and, 113; the livestock industry, 113, 116; Monsanto's Climate Corporation purchase, 76-78, 119. *See also* food industry
agricultural policy, 67-68
agricultural productivity: appropriationism and, 83; increases as rationale for digital agriculture technologies, 124, 131; increases viewed as hunger solution, 54-55, 61, 66-67, 131; productivism and its impacts, 67-68, 200. *See also* Green Revolution
agricultural research and development, 86-89, 123, 136-37
agriculture, alternative, 28, 79, 81, 135-36, 139-40, 179
agriculture, industrial: commodification of farm processes, 83-84; the Salinas Valley, 140-41; the technology treadmill, 70; US farm policy and its impacts, 67-70. *See also* agricultural productivity; agrifood solutions/technologies; digital agriculture technologies; Green Revolution; livestock *entries*
agrifood solutions/technologies: appropriationism and, 83-84, 85, 88-89, 90, 91, 129; the current tech investment landscape, 90-95, 92-93*table*, 97; ecomodernist approaches to, 8; educational institutions and, 1-3, 4-5, 94; efficiency-focused, 58, 91-94, 141, 154-55; farm mechanization and robotics, 69-70, 92-93*table*, 94, 120-21, 128, 132, 156; features

[241]

agrifood solutions *(continued)*
and shortcomings of, 4, 27–28;
food innovation hubs, 32; genetic
engineering, 69, 75–76, 85–86, 94;
goals and rationales for, 83–89; the
influence of Monsanto's Climate
Corporation purchase, 77–78, 119;
as problem-causing agents, 83–86,
99, 129, 135–36; substitutionist
technologies, 84, 85, 88–89, 90,
91, 95. *See also* ag tech; food tech;
Green Revolution
agroecologists and agroecology,
139–40, 153, 154–55, 156
ag tech: in the current investment
landscape, 90, 91, 92–93*table*, 94,
95. *See also* agrifood solutions/
technologies; digital agriculture
technologies; University of
California-Santa Cruz ag tech
initiative
Airbnb, 40, 41, 91
Air Protein, 179–80
AI technologies, 166, 192
alternative proteins, 94, 95, 96–17;
Air Protein, 179–80; animal food
simulacra, 100–103, 107–9, 111–12;
Beyond Meat, 80, 100, 113, 116;
cellular meat, 101, 108, 109,
111–12; and the conventional
livestock industry, 113; enthusiasm and rationales for, 97–99;
environmental claims for, 105,
107–11; health claims for, 105–6,
111–12, 116; the Impossible Burger,
75–76, 100; investments in, 97,
113, 122; market outlook for,
116–17; protein ingredient
powders from novel sources,
103–5, 108, 109; as substitutionist
technologies, 106–12, 121, 122; as
techno-fixes, 100–105, 114–17, 121;
their failure to provide a true
response, 114–16, 134
Amazon, 40–41
animal welfare concerns: alternative
proteins and, 98–99, 105, 107.
*See also* livestock production
concerns
antibiotics, 9, 69, 85, 99, 111, 136. *See
also* herbicides and pesticides
anti-intellectualism: in Silicon
Valley, 184; in university-based
innovation programs, 162, 165,
168, 169, 172–73, 177
antitrust law, 89
Apple Computer, 36, 37, 48
appropriationism and appropriationist technologies: definitions and
features of, 83–84, 199; industrial
agricultural technologies as,
83–84, 85, 88–89; newer agrifood
technologies as, 90, 91, 129–33,
134–35
apps, 10–11, 133
artificial intelligence. *See* AI
technologies
Asia, the Green Revolution in,
59–61
Austin, as tech hub, 32
automation. *See* robotics and
mechanization

Bangladesh famine, 64
Bayh-Dole Act (1980), 146
B Corporations, ix-x, xi, 44, 182
Berman, Elizabeth Popp, 147
Beyond Meat, 80, 100, 113, 116

Biafran famine, 65
big data, digital agriculture technologies and, 130, 134–35
Big Ideas contest, 159–60, 161–63, 164–69. *See also* university-based innovation programs
bioreactors, for alternative proteins production, 109–10
biotechnology, 37–38; agrifood genetic engineering, 69, 75–76, 85–86, 94, 121; UC Berkeley–Novartis partnership, 145–46
Black people, police violence against, 11–12
Blum, Richard, 165
Blum Center on Developing Economies, 165, 174, 176. *See also* Big Ideas contest
Borlaug, Norman, 54–55, 56, 58, 63, 182. *See also* Green Revolution
Boston, as tech hub, 32, 123
Brin, Sergey, 42
Brown, Pat, 142
Bush, George H. W., 21
Bush, George W., 21
business regulation. *See* government regulation

CAFOs (concentrated animal feeding operations), 99, 115–16. *See also* livestock *entries*
Californian ideology, 42, 184
California university system. *See* University of California *entries*
capabilities approach (Sen), 63–66, 199
capitalism, xii, 7, 13, 18, 20; capitalism-friendly origins of the Green Revolution, 58, 59, 182; solution-making and, 182–86. *See also* inequality; social problems
Cargill, 113
cellular agriculture programs, 4, 94
cellular meat, 101, 108, 109, 111–12
certified Benefit Corporations, ix–x, xi, 44, 182
China, 59, 65, 98
Christensen, Clayton, 41
cleantech, 43–44, 51–52, 77
climate change, addressing, 7–8, 11, 114–15. *See also* environmental problems
Climate Corporation, 76–78, 119
Clinton, Bill, 21
Cochrane, Willard, 70
Cole, Teju, 15, 169, 171–72
Collab Fund, 181
colleges. *See* universities and colleges
colonialism, 13, 15, 59. *See also* neocolonialism
Columbus, as tech hub, 32
Commodity Credit Corporation, 67–68
Community Studies major (UCSC), 174–77, 191
competition, 170. *See also* university-based innovation programs
concentrated animal feeding operations (CAFOs), 99, 115–16. *See also* livestock *entries*
context (for solutions/responses). *See* problem framing and analysis
corn, 69, 86
cotton, 86
COVID-19 pandemic and response, 9, 27, 192–93

*Index* [243]

CPGs (consumer packaged goods), 94–95. *See also* innovative foods; processed foods
CRISPR technology, 94
critique: as key to response, 174, 176, 186, 189–91; sidelining of, 48, 155, 186
cultured meat. *See* cellular meat
Curtis, Lisa, ix, x–xi, xiv, 14, 168. *See also* Kuli Kuli

data digitalization: apps for, 10–11, 133. *See also* digital agriculture technologies
DDT, 69
Desai, Mihir, 186
design thinking, 16–18, 168, 199
development, 12–13. *See also* Green Revolution
*Diamond v. Chakrabarty* ruling, 37
*Diet for a Small Planet* (Lappé), 108–9, 117
digital agriculture technologies, 95, 118–37; advent of, 122–23; as appropriationist, 129–33, 134–35; Climate Corporation's climate change-monitoring technology, 77; concerning implications of, 129–30, 131–33; enthusiasm and rationales for, 118–21, 122–24, 125–26, 131–32; market outlook for, 134–35; as solutionist, 121, 124–29, 133; their failure to address actual needs, 132–37
disruptive innovation, 41, 43; Silicon Valley food tech solutions seen as, 79–81, 82, 123
Doerr, John, 43–44
DuPont Company, 7

Easterbrook, Gregg, 54–55, 71
ecomodernism, 7–8, 199
economic inequality. *See* inequality
educational institutions. *See* universities and colleges
education otherwise, 177
efficiency, as problem frame/solution focus, 42–43, 58, 91–94, 137, 141, 154–55
eGrocery investment, 90, 91, 92–93*table*, 95
Egypt, 59
endowments, in Sen's hunger framework, 63, 199
engineering fixes. *See* techno-fixes
engineers and engineering, 149–50, 153–54; geoengineering, 7, 8, 114–15
entitlements, in Sen's hunger framework, 63–64, 199
entrepreneurship: the entrepreneurial university, 147; valorization of, 24, 36, 184, 187
environmental optimism, 71–73, 114
environmental problems and rationales: addressing climate change, 7–8, 11, 114–15; agrifood technologies and, 85; cleantech as solution, 43–44, 51–52; digital agriculture technologies and, 124, 131–32; environmental claims for alternative proteins, 105, 107–11; environmental improvement as Silicon Valley goal, 43–44; envisioning effective responses to on-farm environmental problems, 135–36; Malthusian framings of, 57–58, 71–72; the tech sector's

failure to address problems, 192–93. *See also* livestock production concerns
Enviropig, 86
Epstein, Jeffrey, 3
*Essay on the Principle of Population* (Malthus), 56
European tech hubs, 32
expertise-driven solutions, 13–14, 15, 55, 126. *See also* rendering technical

Facebook/Meta, 30, 42
famines, 59, 60, 63, 64–65. *See also* hunger
farm equipment, 69, 91, 92–93*table*, 94, 156
farm labor. *See* labor
federal funding. *See* government investment/funding
federal policy. *See* government regulation; public policy
fertilizer use, 83–84, 85
Food and Agribusiness Innovation Prize (MIT), 164
food assistance programs, 24, 61, 66, 72
FoodBytes!, vii–viii, x
food distribution, tech innovations in, 90, 91, 92–93*table*
food industry, vii, 4, 88, 106–7. *See also* agribusiness; processed foods
food security, 66; protein supply concerns, 98, 99, 100, 103, 104. *See also* Green Revolution; hunger
food system problems: the social nature of, 79, 86, 87–88, 153; solutionism and, 10–11, 27; ubiquity and complexity of, 78–79. *See also* food security; hunger

food tech, 74–95; the current investment landscape, 90–95, 92–93*table*, 97; as disruptive innovation, 79–81, 82; the OpenAg personal food computer, 2, 3–4, 11; origins of, 76–81; Silicon Valley's enthusiasm for, 74–76, 77–81, 122; as substitutionist, 84; traditional agribusiness and, 76–78, 88–89; venture capital investment in, 80–81, 90–95, 92–93*table*. *See also* agrifood solutions/technologies; alternative proteins; digital agriculture technologies
Ford Foundation, 59, 60, 62
Founders Fund, 76
Freire, Paulo, 173
Friedland, William, 140
Friedman, Milton, 20
Friedman Food & Nutrition Innovation Prize (Tufts University), 164
*From Farming to Biotechnology* (Goodman, Sorj, and Wilkinson), 83–84
Fukushima nuclear disaster, 2
funding and its impacts: academic funding pressures and sources, 25–26, 88, 89, 123, 141, 145–46, 147, 148–49; venture capital constraints on problem response, 51–52. *See also* government investment/funding; philanthropy; venture capital

Genentech, 38
genetic engineering, 69, 75–76, 85–86, 94
geoengineering, 7, 8, 114–15

Gilmore, Ruth, 190–91
Giridharadas, Anand, 185
global health initiatives, 14–15
Global Poverty and Practice minor (UC Berkeley), 176–77
glyphosate, 108
GMOs, 69, 75–76, 85–86, 94
Golden Rice, 86
Goldstein, Jesse, 51–52, 200
Goodman, David, 83–84
Google "Don't be evil" motto, 30, 42
Gore, Al, *An Inconvenient Truth*, 43
Gorz, Andre, 190
government investment/funding: patentability and, 146; research funding, 88, 146, 148; and Silicon Valley's early history, 33–35, 39; the Small Business Investment Company and the rise of venture capital, 39; social welfare programs, 21, 22, 24, 61, 66, 72, 144; state support for California's universities and colleges, 143, 144–45, 147
government regulation (of business): early twentieth-century expansion of, 20; hands-off regulatory policy and the tech industry, 37–39, 184; neoliberalism and, 21, 22. *See also* neoliberalism; public policy
green capitalism, 44. *See also* environmental problems and rationales
Green Revolution, 53, 54–62, 182, 185; context, origins, and outcomes of, 58–61, 62, 70–71; critiques and detractors, 55, 60–62; hunger problem framing and its conceptual context, 55, 56–57
green tech. *See* cleantech

Hajer, Maarten, 11, 200
Harper, Caleb, 2
Harvard University, 32, 34
Hayek, Friedrich von, 20
health claims and concerns. *See* animal welfare concerns; human health
Hedberg, Russell, 102–3
herbicides and pesticides, 69, 76–77, 85–86, 108, 135, 136; digital agriculture technologies and, 131–32
Hewlett, William, 33
Hewlett-Packard, 33–34
Holmes, Elizabeth, 3, 48
Homebrew Computer Club, 36
homelessness, 25
*How to Do Nothing: Resisting the Attention Economy* (Odell), 186–87
hubris, 32, 158, 162, 183, 189
human health: agrifood technologies and human health concerns, 85, 99; global health initiatives, 14–15; health claims for alternative proteins, 105–6, 111–12, 116
humanities expertise, undervaluing/marginalization of, 145, 149, 150–51
human labor. *See* labor
hunger and food insecurity: broadly-conceived responses to, 62–66, 72; during the American Great Depression, 67; vs. food security, 66; Malthusian notions of, 56–57, 66–67, 71–72, 104, 131. *See also* famine; Green Revolution

impact imperative: in academia, 148–51, 157–58; in the UC Berkeley Big Ideas contest, 165–66
impact investing, 44
Impossible Foods/Impossible Burger, 75–76, 100, 116
*An Inconvenient Truth* (film), 43
incubators, 46
India: Green Revolution and, 59, 60, 61; hunger and food distribution in, 61, 63, 64, 66; rising meat demand in, 98
Indigo Agriculture, 80
Indonesia, 59
industrial agriculture. *See* agriculture, industrial
inequality, 20; avoidance/dearth of real responses to, 8, 14, 18–19, 71–72, 186; entrepreneurialism viewed as solution to, 24; neoliberal policies and, 22; the UC Berkeley Global Poverty and Practice minor, 176
innovation: the impact imperative, 148–51; in the Silicon Valley ecosystem, 36–37, 45. *See also* disruptive innovation; university-based innovation programs
innovation competitions. *See* university-based innovation programs
innovative foods, 91, 92–93*table*, 94–95, 97. *See also* alternative proteins
Intel, 35
intensification, 68, 69, 71, 199. *See also* agricultural productivity
interdisciplinarity, 150

International Rice Research Institute (IRRI), 59–60
internet development, 38, 40
Irish Famine, 65
IRRI (International Rice Research Institute), 59–60

Jobs, Steve, 36, 48

Kellogg's, ix–x, xi
Kerr, Clark, 142
Klein, Naomi, 8, 114
Kuli Kuli, viii–xii, xiii–xiv, 24, 168

labor: the advent of neoliberalism and, 20, 21; considering the impacts of new technologies on, 166; farm labor in California, 87–88, 128–29; labor power as endowment, in Sen's analysis, 63, 64; labor-saving agricultural technologies, 69–70, 83
labor unions, 21, 174
land-grant educational institutions, 86–88, 138
land reform, 58, 72, 185
Lappé, Frances Moore, *Diet for a Small Planet*, 108–9, 117
LinkedIn, 76
Li, Tania Murray, 12–14, 200, 201
livestock industry, 69, 113, 116
livestock production concerns, 78, 86; alternative proteins and, 95, 97, 98–99, 102–3, 105, 107–9; envisioning effective responses to, 114–17

malnutrition. *See* hunger and food insecurity

Malthus, Thomas, 56–57, 66–67. *See also* neo-Malthusianism
Mann, Charles, 71
margarine, 106
Massachusetts Institute of Technology. *See* MIT
meat substitutes. *See* alternative proteins; *specific products*
mechanization. *See* robotics and mechanization
Memphis Meats, 113
Meta/Facebook, 30, 42
Mexico, the Green Revolution and, 58–59, 61
microfinance initiatives, 24
Microsoft, 32
military contracting, in Silicon Valley's early history, 33–35
Mirowski, Philip, 147
MIT, 32, 33; food and agriculture initiatives and competitions, 1–4, 10–11, 26, 164
Monsanto, 76–77, 86; the Climate Corporation purchase, 76–78, 119
moonshots, 178–80, 199
Moonshot Snacks, 179, 180
moringa, viii–ix. *See also* Kuli Kuli
Morozov, Evgeny, 9–10, 50, 200
Morrill Act, 87
Musk, Elon, 44

Naess, Arne, 8
National Science Foundation, 148, 149
National Semiconductor, 35
neocolonialism, xi, 15, 163. *See also* will to improve
neoliberalism and its impacts, 19–26, 200; educational institutions and, 25–26, 141–47, 170; vs. government role in Silicon Valley's origins, 34–35; historical overview, 20–22
neo-Malthusianism, 66–67, 71–72, 104, 131, 200
New Left, and the personal computing revolution, 35–36
Non Aligned Movement, 59
nondisruptive disruption, 51, 200
nonprofit sector programs and activities, 24–25. *See also* philanthropy
nonreformist reform, 190–91
Novartis Corporation, 145–46
novel farming systems, 90, 91, 92–93*table*, 95
nuclear energy, 8
nutritional claims, for alternative proteins, 111–12

Oak Ridge National Laboratories, 7
Odell, Jenny, 186–87
olestra, 112
O'Mara, Margaret, 32–33
online food sales and delivery technologies, 90, 91, 92–93*table*, 95
OpenAg Initiative (MIT), 1–4, 10–11, 26
organic agriculture, 28, 79, 136, 139
organizing and activism, 5, 172, 173–77, 185–86; the challenges of food activism, 78–79; strategy as key to, 189–90; UC Berkeley Global Poverty and Practice minor, 176–77; UCSC Community Studies program, 174–77, 191. *See also* responses

Packard, David, 33
Page, Larry, 42. *See also* Google

Paquet, Gilles, 18
Patagonia, 179
patents and patentability, 37–38, 47, 49, 112, 146, 147
paternalism. *See* will to improve
PayPal, 44, 76
pea protein isolates, 109
pedagogy: education otherwise, 177; praxis-based, 172–77, 191. *See also* university-based innovation programs
Perdue, 113
performativity, 48. 49. 147, 167, 168, 173, 182, 189
personal computing, 36, 37
pesticides and herbicides, 69, 76–77, 85–86, 108, 135, 136; digital agriculture technologies and, 131–32
philanthropy, 23, 24, 25, 26
Philippines, the Green Revolution in, 59–60
pitching and pitch culture, 30, 48–50, 52, 168; in university-based innovation competitions, 161, 162, 164, 167–68
plant-based protein products, 100, 108–9. *See also* alternative proteins
police violence, 11–12
political problems. *See* social problems
politics, avoidance of, 7, 14, 17, 25, 72, 185
population growth, hunger and, 56–57, 71–72
praxis-based pedagogies, 173–76, 191
precision agriculture technologies. *See* digital agriculture technologies

private sector/private industry: agricultural research and development and, 88–89, 123, 140–41; neoliberalism and the increased role of, 23–24, 25–26, 34–35; ties to educational institutions, 26, 140–41, 142, 145–46, 165. *See also* agribusiness; food industry
probiotics, 115, 135, 136
problematization, 13–14, 200
problem closure, 11, 200
problem framing and analysis: for alternative protein solutions, 98–99, 102–3, 104, 114; design thinking approaches to, 17–18; for digital agriculture solutions, 124–26; by Green Revolution promoters, 55, 57–58, 59; reductive problem definitions/analyses, 6, 8, 11–12, 14, 153–54; Sen's framing of the hunger problem, 63–66; in Silicon Valley pitch culture, 49, 52; solutionism and, 9–11, 15, 18; techno-fixes and, 7–9, 15, 19; for the UCSC ag tech program, 152–55; in university-based innovation competitions, 162–63, 166–67, 168, 169; the will to improve and, 13–14. *See also* responses; solutionism; solutions, misconstrued; techno-fixes
processed foods, 85, 106–7; alternative proteins as, 108–10, 112. *See also* food industry; innovative foods
productivism, 67–68, 200
productivity. *See* agricultural productivity; Green Revolution

Proposition 13 (California), 144
protein foods, alternative. *See* alternative proteins
protein supply concerns, 98, 99, 100, 103, 104
public policy: the Morrill Act and origins of land-grant institutions, 87; neoliberalism in, 20–22, 27, 37, 143–47; regulatory state expansion in the twentieth century, 20; social welfare programs, 21, 22, 24, 61, 66, 72, 144; tax policy and its impacts, 21–22, 24–25, 38, 39, 144–45; US agricultural policy and its impacts, 67–70. *See also* government regulation; neoliberalism

racial capitalism, xi–xii, 189
radical optimism, xiv, 154, 155
Reagan, Ronald, 21, 143; Reagan-era neoliberalism and the University of California system, 143–47
recycling, 110–11
reform, nonreformist vs. reformist, 190–91
regenerative agriculture, 79, 81, 179
rendering technical, 13, 14, 55, 200
research expertise, support, and funding, 86–89, 123, 136–37, 145–46, 148–49
responses: broadly-conceived responses to hunger and food insecurity, 62–66, 72; defined, 200; envisioning effective responses to agriculture-related problems, 114–17, 135–36, 187–88; vs. solutions, 18–19, 27–28, 153–55, 187–91. *See also* organizing and activism

responsible innovation, 16, 18, 151, 200
rice, 59–60, 86
risk and risk-taking, 36–37, 39
robotics and mechanization: enthusiasm for, 75, 120–21, 127–28, 181, 192; impacts and implications of, 69–70, 132, 156, 166; investment in agrifood robotics and mechanization technologies, 92–93*table*, 94. *See also* digital agriculture technologies
Rockefeller Foundation, 58, 59, 60, 62
rural sociology, 87–88

Salinas Valley, 140–41
San Francisco Bay Area: Bay Area counterculture and its Silicon Valley impacts, 35–36, 42; as tech hub, 32, 40. *See also* Silicon Valley
Saxenian, Anna, 36
SBIC (Small Business Investment Company), 39
SBIR (Small Business Innovation Research) program, 39
*Science-Mart* (Mirowski), 147
Seattle, as tech hub, 32
semiconductor industry, 35
Sen, Amartya, 62–66, 72, 105, 199
Silicon Valley, 22; geographic setting, 31–32; Silicon Valley–UCSC relationship/partnerships, 140
Silicon Valley agrifood tech. *See* agrifood solutions/technologies; ag tech; food tech; *specific technologies and products*
Silicon Valley ecosystem and ethic, 29–31, 32–33, 45–50, 184–85, 189;

the allure of food tech, 74–76, 77–81, 122; incubators and accelerators, 46–47; innovation, change, and risk-taking, 36–37, 41, 43, 45, 51; moralizing impulses and social-benefit efforts, 30, 41–44, 51–52; pitching and pitch culture, 30, 47, 48–50, 52; replicated in academia, 158, 159–64; solution-making as feature of, 30–31, 44–45, 50–51; techno-optimism in, 72–73; venture capital and its influence, 38–39, 47, 51–52

Silicon Valley history, 33–45, 183–84; internet and e-commerce development, 38, 40–41; origins in the war industry, 33–35; personal computers and the shift to consumer-focused innovation, 35–37; the regulatory environment, 37–39, 184; the rise of social-benefit goals and investments, 41–44; the rise of venture capital, 38–39

*Silicon Valley* (TV series), 29–30, 48

Small Business Innovation Research (SBIR) program, 39

Small Business Investment Act, 39

Small Business Investment Company (SBIC), 39

social change process: educating/involving students in, 174–77. *See also* organizing and activism; responses

social enterprise/social-benefit capitalism, ix–xi, 23, 44; Silicon Valley tech companies and, 42–44, 51–52

socially responsible investing, 44

social problems: food system problems as, 79, 86, 87–88, 153; the tech sector's failure to address, 192–93. *See also* food system problems; inequality; problem framing and analysis; responses; social change process

social science expertise, undervaluing/marginalization of, 87, 145, 148–51, 154, 155–58

social welfare programs, 21, 22, 24, 144; food assistance, 24, 61, 66, 72

solutionism and solutionist technologies, 6, 15, 19; alternative proteins as, 95, 100, 102–4; definitions and features of, 9–12, 18, 183, 200; digital agriculture technologies as, 121, 124–29, 133; the Green Revolution and, 55; university-based innovation competitions and, 162. *See also* problem framing and analysis

solutions, misconstrued, 4, 183, 185, 192–93; for alternative protein products, 114–16, 134; for digital agriculture technologies, 132–37; reductive problem definitions/analyses, 6, 8, 11–12, 14, 153–54; and university-based innovation competitions, 169–73. *See also* solutionism; techno-fixes; techno-optimism

solutions and solution-making: allure of, xiv; neoliberal roots of the solutions imperative, 19–26, 183–84; refusing the solutions imperative, 187; vs. responses, 18–19, 27–28; shortcomings of, solutions and solution-making *(continued)* xiii–xiv, 15–16, 26–28, 180, 182–86, 192–93; in Silicon Valley culture, 30–31, 44–45, 50–51; the solutionism context, 6, 9–12, 15, 18, 19; solutions defined, xiii, 6, 200; the techno-fix context, 6–9, 15, 19; the will to improve context, 6, 12–16, 17–18, 19. *See also* responses

Sorj, Bernardo, 83–84

soybeans, 108

Soylent (beverage product), 97, 108

*Soylent Green* (film), 96–97

Stanford University, 31, 33, 34, 88, 142

start-up culture, 36–37, 38–40, 45–50. *See also* Silicon Valley ecosystem and ethic

STEM fields, elevation of, 26, 141, 145, 146–47, 149–51, 153–55

strategy, 19, 189–90

substitutionism and substitutionist technologies: agrifood technologies as, 90, 91, 95, 106–12, 121, 122; definitions and features of, 84, 88–89, 200; lack of transparency and, 84, 108, 112; processed foods as, 106–7

sugar, 106

supply chain management technologies, 90, 121, 125–26

tax policy and its impacts, 21–22, 24–25, 38, 39, 144–45

techno-fixes, 183, 188, 201; alternative proteins as, 100–105, 114–17, 121; features and limits of, 6–9, 15, 19, 72; geoengineering, 7, 8, 114–15; the Green Revolution as,

55, 57–58, 62; the impact imperative and, 157–58; promotion by university programs, 157–58, 162–63, 171. *See also* problem framing and analysis; techno-optimism; university-based innovation programs

technology treadmill (in agriculture), 70

techno-optimism, 71–73, 114, 181, 201

tech sector/tech industry: development and economic influence of, 22–23, 26; its failure to address our biggest challenges, 192–93; San Francisco Bay Area as tech magnet, 32, 40. *See also* Silicon Valley *entries*

TED and TEDx Talks, 43–44, 48–49

Teller, Eric, 178

Terman, Frederick, 33, 34

Tesla Motors, 44

Thatcher, Margaret, 21

Theranos, 3, 48

Thiel, Peter, 76

third world, as term, 59

transparency/lack of transparency, 84, 97, 108, 112, 126–27

Trump, Donald, 21

Tufts University Friedman Food & Nutrition Innovation Prize, 164

Tyson, 113

Uber, 40, 41, 43, 93

UCSC. *See* University of California–Santa Cruz

underlying problems. *See* problem framing and analysis; responses; social problems; solutions, misconstrued

United Farm Workers, 174
universities and colleges: agrifood technology programs at, 1–3, 4–5, 94; the elevation of STEM fields and its impacts, 26, 141, 145, 146–47, 149–51, 153–55; funding pressures and sources, 25–26, 88, 89, 123, 141, 145–46, 147, 148–49; the impact imperative, 148–51, 157–58; the impacts of neoliberalism on, 25–26, 141–47; interdisciplinarity, 150; praxis-based pedagogies, 173–76, 191; the rise of solution-making pedagogies, 160–64; as the seats of traditional agriculture and food science research, 86–89; tech industry development and, 32, 38, 40; ties to the private sector, 26, 140–41, 142, 145–46, 165. *See also* University of California *entries*; *other specific schools*
university-based innovation programs, 3–5, 159–77; the exclusion of broader knowledge and critical analysis, 160, 163, 166, 168, 172–73; the focus on solutions and techno-fixes, 5, 157–58, 161–63, 171; the impact imperative in, 165–66; the pitch-culture aspect of, 161, 162, 164, 167–68; problem framing/analysis in, 162–63, 166–67, 168, 169; proliferation of, 4–5, 160–61; student preparation/training/mentoring and, 163–64, 167–68; the UC Berkeley Big Ideas contest, 159–60, 161–63, 164–69; undesirable effects of, 4, 5, 169–73

University of California–Berkeley, 138, 145–46; Big Ideas contest, 159–60, 161–63, 164–69; Global Poverty and Practice minor, 176–77
University of California–Davis, 94, 138
University of California–Santa Cruz, 5, 138–40, 145, 147; Community Studies program, 174–77, 191
University of California–Santa Cruz ag tech initiative development, 136–37, 138–58; academic context, 148–51; the campus community's divergent/contentious views of, 152–57; historical context, 138–40, 142–47; motivations for, 140–41
University of California system, Reagan-era neoliberalism's impacts on, 143–47
USDA, the Green Revolution and, 58

veganism, 100, 101
venture capital, 38–39, 47, 51–52; agribusiness venture capital programs, 113; agrifood tech investment, 80–81, 90–95, 92–93*table*, 97
vertical farms, 90, 91
Vietnam, 59
Vinsel, Lee, 17–18

waste production and processing, 98, 99, 109, 110–11, 181
Weinberg, Alvin, 7
White Savior Complex, 15, 169, 170–72, 201

Wilkinson, John, 83–84
will to improve, 188–89, 201; alternative proteins and, 102; features and shortcomings of, 6, 12–16, 17–18, 19; the Green Revolution and, 55, 62; moralizing impulses in Silicon Valley, 30, 41–44, 51–52; and the rise of social entrepreneurship, 23–24; university-based innovation competitions and, 163, 169, 170–72; White Savior Complex, 15, 169, 170–72, 201

World War II, Silicon Valley's war industry origins, 33–35

Wozniak, Steve, 36

X, 178

Yong, Ed, 9

Zuckerberg, Mark, 30, 42

Founded in 1893,
UNIVERSITY OF CALIFORNIA PRESS
publishes bold, progressive books and journals
on topics in the arts, humanities, social sciences,
and natural sciences—with a focus on social
justice issues—that inspire thought and action
among readers worldwide.

The UC PRESS FOUNDATION
raises funds to uphold the press's vital role
as an independent, nonprofit publisher, and
receives philanthropic support from a wide
range of individuals and institutions—and from
committed readers like you. To learn more, visit
ucpress.edu/supportus.